WHEN THE UNIFORM COMES OFF

WHEN THE UNIFORM COMES OFF

Navigating the Rough Seas of a Military Marriage

Eric L. Allen

Final Step publishing

Suffolk, Virginia

When the Uniform Comes Off:
Navigating the Rough Seas of Military Marriage

Copyright © 2022 by Eric L. Allen

Final Step Publishing, LLC
PO Box 1441
Suffolk, VA 23439
www.finalsteppublisher.com

Paperback ISBN: 979-8-9850005-1-1

Cover design by Cooke Classic Branding & Design | www.cookeclassic.com

Cover Image by Mike Dragon with Dragon Studio

For Worldwide Distribution. Printed in U.S.A.

DEDICATION

To my wife, Ellasin, your love, patience, and support, fuels my creativity and has helped me become a better man. To my children Chalanna, Darius, and Ayshia you were my driving force, and my inspiration for success.

FOREWORD

I have been close friends with Eric for fifteen years. We first met while serving together onboard the USS Theodore Roosevelt. I will never forget our first interaction. He asked, "Where are you from?" and I replied "Kentucky." He was from Kentucky also. His next question would shape our relationship forever. He asked, "Are you red or blue?" to which I replied "Blue." His answer to the question was red, referring to the University of Louisville; my response of blue referred to the University of Kentucky. This would start a lifelong rival that grew our relationship even stronger.

To say Eric changed my life is an understatement. Not only did I learn the mechanics of running a fuel system, but he also renewed my faith in God. I was going through a rough time in my personal life, not knowing Eric would help me turn it all around. Anytime we encountered a stressful situation or hurdle, he would always say let's refer to the sword, meaning his Bible. No matter the situation, it always turned out to be okay. I then realized he possessed faith like no other person I have ever met. I have personally witnessed

Eric always take time out of his busy day to help a young sailor. Whether it be advice on their career or personal battles, he was always there to assist them.

If anyone is qualified to write such a powerful book, it is Eric—a person who can command a room full of a thousand people and not bat an eye. So why is it that a man with such stature cannot do the same while he is at home? When Eric started discussing the book with me, a light bulb immediately went off. I suffered from the same issues as Eric. I can lead 800 men and women, spit out orders, and mentor them, but I could not have a simple conversation with my wife. I would carry my Navy work demeanor home without even realizing it. This was causing great strain at home, and I couldn't turn it off. Through several phone conversations with Eric, he helped me understand how to be me at home and not treat my spouse like one of my sailors. I am forever grateful for his expertise and mentorship throughout the years.

The struggle to leave the uniform at work is real. I know of many people who cannot turn off being a leader of men and women when they are at home. Eric helps us understand the root of the issue and gives sound advice on how to make the change. If you struggle when the uniform comes off, then this is the book you have been waiting for. Get ready to see yourself in an entirely different light and learn how to be you.

—**Chris Jones**
LCDR USN

CONTENTS

INTRODUCTION

I AM CONVINCED THAT being married and staying married, especially today, is harder than at any other time throughout history. Marriage has lost the appeal that once made it sacred in our society particularly amongst many young people. It is rapidly transitioning from being the grand prize to merely being seen as an item on the drive-thru menu of the fast-food restaurant some see as life. We live in a sex-crazed nation that promotes having multiple partners and material wealth; this demand leads to more time and attention to the work environment to obtain the things society tells us we need to be happy. With our country's moral compass rapidly declining, it is easy to see why the divorce rate climbs and many marriages continue to fail, some soon after consummation. Add to the mix increasing military mission demands and the arduous duty requirements of a military service man or woman, and you now have a successful recipe for marital disaster.

God intended for marriage to have meaning, worth, value, and to be a symbol of His relationship with His people. He designed marriage as a partnership to be fruitful, not only in the manner of reproduction, but also in creativity.

"God blessed them and said to them, "Be fruitful, and increase in number, fill the earth, and subdue it. Rule over the fish in the sea, and the birds in the sky and over every living creature that moves on the ground." –Genesis 1:28

Yes, we are to be productive and reproduce to ensure the existence of humanity, however, we are also to be fruitful and create new ideas, bring healing to hurting hearts, and spread love instead of hate; in other words, to jointly be representatives of God in the world by impacting other people's lives and leaving behind a legacy for your children. This is God's plan for marriage but notice in the scripture when He gave the commandment to be fruitful and multiply, He first had to bless them. That is because before you can influence the world as a couple, you must be blessed by God to be productive, fruitful, and endure the hardships of life.

When we do things as a couple without the blessings of God, we can find ourselves in desolate places and unable to produce fruit as a couple; in fact, desolate places rarely produce any fruit at all. There is scarce vegetation in desolate places, it is often very dry and accompanied by thorn bushes, thistles, and the carcasses of dead animals that perished there. This is evident when those who have been married fifteen, twenty, thirty, and even forty years finally conclude to abandon a stagnant, unfulfilling, barren marriage that produces nothing more than the thorns and thistles of bitterness and resentment. Many find themselves financially well off and quite successful in their careers and businesses, yet their life together has been reduced to a dead carcass within the most important union God has instituted on earth—marriage. However, I believe some of the greatest challenges and obstacles to overcome for married couples are those which accompany the lifestyle and career of the men and women who have chosen to join the United States Military.

Divorce is an industry. Contested divorce in America accounts for between $50 and $175 billion dollars annually, but when you look at the military family, the statistics are alarming. A career website by the name of Zippia[1] examined professions with the highest divorce rates among people under thirty. Three of the top ten professions with the highest divorce rates came from military personnel. First line supervisors had the highest divorce rate of 30 percent. These are the individuals leading troops and supervising enlisted service members performing military operations. Air weapons and tactical operations came in at number four at over 15 percent and any service member with an unspecified rank came in at seventh overall and just under 15 percent.

Keep in mind these are statistics for members under thirty. This does not consider those who get divorced after a twenty- or thirty-year obligation of service. It also does not account for those who have been on active duty more than twelve to fifteen years where the divorce rate drops to 3-3.5 percent overall. Therefore, according to the statistics, those dealing with the bulk of divorce in the military are under thirty years of age and serving on active duty for under ten years of service. But that doesn't mean those that are older are happy with how their marriage is either; many have learned to just deal with being married so they do not have to give up half of their retirement pay in spousal support. Many do not want to struggle with what to tell the children or wrestle with how to divide up the property or possessions afterwards.

1 https://www.zippia.com/advice/divorce-rates-job-industry/

Service members must overcome a myriad of problems to maintain a healthy marriage. I think the same can be said for our country's first responders and health care workers or any profession that calls its members into a form of civil duty that requires most of their time and demands them to selflessly place others before themselves. They endure some of the same struggles that regular marriages do such as arguments over bills and financial struggles, disagreements about the children, and problems with the in-laws. However, throw in stressful mission demands at work, separation for extended periods of time, the fear of harm or danger, a narcissistic attitude that so often comes with a military mindset, and there is a whole new dynamic most of the country knows absolutely nothing about.

Did you know that 93 percent of the nation's population has never been in the armed forces, has never gone to boot camp, and never been on deployment? While it is true that less than 0.04 percent of the country is on active duty at one time, when you consider our veterans, there is only about 7% of the population who has EVER served in the United States Military.

Therefore, the military family, while honorable, is most often flawed and damaged by the demands of the service it provides to their country. Additionally, the length of separation easily drives a wedge in the marital relationships of service members. As a young sailor, I was disgruntled as many young sailors are. Without the privilege of knowing the big picture, I could not seem to fathom how my minuscule duties of shining brass, sanding rust, and painting passageways had anything to do with national security. I detested the fact that the moment something as small as being a few minutes late or performing a task incorrectly instantly became a global event. I was not allowed to voice my displeasure about anything, my words carried no weight because I had no rank, and I very much disliked being in the Navy as a result. I began looking forward to the time I would get out of the Navy and go back to living a normal life, to get hired at Ambrake or DANA, two big automotive parts factories that were in my hometown. I wanted to grow a beard and be my own person again. I longed to go back to what was familiar because everything became so foreign to me.

However, what I was trying to run away from became a place of solace for me as time went on. There is a lot of time to think when out to sea; in the middle of the chaotic orchestra of aircraft carrier flight deck operations is a sense of peace where the sky meets the water. The medical benefits and housing pay I was receiving for having dependents motivated me to succeed. I began to look forward to getting underway as it became a break from arguing with my wife or raising the children. It might have been extremely dangerous at times, but it was just me, and this made me selfish.

Unfortunately, somewhere between the longing to get out and taking on a different occupation and the numerous underway periods I was engaged in, I changed. I took on a different personality, and I became the Navy. The change was subtle—it slowly infiltrated my thinking and eventually became tangible in my behaviors. I was by no means a gung-ho, "Joe Navy" type of person. I often tiptoed in an area between naval regulations and what I wanted to do. I pushed against the rules but not enough to be in trouble. But in my ambitions to advance, to be successful, and be a better provider for my family, I slowly began to enforce the very regulations I used to push against.

As I began to advance up the ranks, I became more calloused, more irritable, and less compassionate. My wife, Ellasin, would bring these minor changes to my attention because she noticed a difference in how I talked to her and our kids. But I chose to ignore her complaints because I did not know how to take the uniform off. I was becoming more aggressive, more arrogant, and more prideful. The military, like any demanding occupation, has its own culture, language, and ideology. Too many times the belief is that to be successful one must take on the culture, talk the talk, and subscribe to the politics of the field. Simply put, I learned the rules to the game, and I played it well.

What happens when you realize you have changed as a person but cannot put your finger on when the change happened? The change, while it is beneficial in your professional life, proves to be damaging in your personal one. The change fuels arguments and heated disagreements until eventually a seed of resentment is sown and a monster begins to grow behind the scenes. This beast that is growing is doing so behind the mask that many of our military men and women wear, behind the image they must portray to their loved ones to prevent concern or hide inconsistencies in their behaviors and marital obligations. Growing behind the lies they tell themselves as they continue to deny any problems. It is getting stronger from the perseverance to stay in a career that has damaged the very family it provides for. Once the uniform comes off and the service member is forced to cohabitate with the one(s) they love every day on a continual basis, the beast that has been growing during the times of separation begins to rear its ugly head.

It is a demon that we are all familiar with and all too many of our brothers and sisters in arms have been victims to in their careers. Many prominent doctors, firefighters, police officers, and clergy have fallen prey to this beast as well. It is one that destroys the family and negatively impacts everybody connected to the individuals involved. It grows strong in the corners of low self-esteem and depression; it garners its energy from a lack of attention, a withdrawal of emotion, or the sense of no respect. It breeds bitterness, anger, and mistrust all while destroying hearts, hope, and the future of promise and destiny. It is the spirit of divorce.

As a Navy man for over twenty-four years, I have spent more of my life with the Navy than I have with my own family. The picture I found myself portraying was one of a person my wife and children knew as a loving father, a loyal husband, and Christian minister. It was of a person I so desperately wanted to be but sadly no longer was. The scary thing is I could not remember exactly when I lost me. I hope that the revelation of my own mistakes and the devastating consequences it caused will help military couples survive and their families to heal and prosper.

Through much prayer and patience, my beautiful wife and I have survived what I call the "triple threat"— trials, tribulations, and troubles. It was by God's grace that we continue to love each other and are happier than we have ever been with one another. You, too, can have this, but please understand it takes work. If your relationship is to survive and you want to be happy with your spouse, it is something you BOTH must want, and you BOTH must work at. Therefore, through our transparency, we labored together to produce this book and have endeavored to help not only military couples, but all marriages that struggle through the rough seas of life. While this is not the ultimate authority on a successful military marriage, we pray that you let it be a tool you can use to shape and mold your marriage into the union we believe God intended it to be.

If you are a military member finding your marriage to be in a desolate place and one lacking life and vitality, this book will help you understand why and how to redirect your thinking to achieve a positive relationship where all members of the family grow and benefit from the relationship they share with one another. In fact, if you are not a military member but your career is one that is life threatening or of great stress with extremely high demands, you, too, may deal with some of the things military families struggle with. Hence, this book is for you and your family as well.

Chapter 1
Duty Calls

"I reluctantly pulled into the driveway of my house and just sat there, unable to get out of my truck and go into the house. I sat there because I was feeling something I've never felt and was not aware of it until now. I don't want to be here anymore!"

I NEVER THOUGHT I would be in the position I found myself in on that day, sitting in the driveway in my truck after an argument with my wife in which I felt our marriage was over. After fourteen years of marriage and almost fifteen years in the Navy, I could not be sure how I arrived at this position, but the problem could have been prevented if I had known what to look for when we first got married. One of the most important times a man and a woman can share together is the first year of marriage. It is during this time that a couple begins to unpack each other emotionally, physically, intellectually, and spiritually. They spend every waking minute together; they go everywhere together, cook for each other, and appear to be inseparable. There are things you learn about each other during this time, and a bond is created that cannot be duplicated through phone conversations, text messages, or social media. God designed it this way and is so serious about a strong, healthy marriage that He has given specific instructions in the Old Testament concerning the first year of marriage.

"If a man has recently married, he must not be sent to war or have any other duty laid on him. For one year he is to be free to stay at home and bring happiness to the wife he has married."
—Deuteronomy 24:5

Notice how God says after a man takes a wife, he should not go out to war or be charged with any business for one year. He is to stay home and minister to his new wife, or in other words, to rejoice in the new marriage with her and do things to make her happy. Unfortunately, the modern military man or woman does not have this luxury in the United States. If the service member gets married after they have already joined the prestigious ranks of the United States Armed Forces, unfortunately for them, they may have unknowingly begun their marriage at a disadvantage. Quite frankly there is probably no job in today's culture that is going to allow a person to take a year off just to be with their new spouse, but according to the Word of God, I am led to believe if they did there would be fewer affairs, fewer counseling sessions, and a lot less money spent on divorces.

There is a reason God said to refrain from going to war first. When I got out of boot camp, I married my pretty little wife and left two weeks later for an eight-month deployment. I excelled in basic training: I was an exceptionally good athlete in high school; therefore, I managed the physical conditioning better than most. What I came to discover is that the greater challenge is separation. Separation from home and from everything that is familiar proves to be a greater challenge than the physical requirements.

I was sent to Recruit Training Command in Great Lakes, Illinois for basic training. Now I am from Kentucky and being close to Fort Knox, the Amy was the only military to which I was ever exposed. And here I am enlisting in the Navy, and I was not familiar with anything about that branch of service. I was a decent student and smarter than my grades indicated. I did not like school but tolerated it to remain eligible to play sports.

When my high school football days were over, I attended Kentucky State University for a semester but had to leave school due to campus violations. I was going nowhere fast and needed a way out if I were to make anything of myself. When we found out that my wife was expecting, I knew I could not continue this reckless journey. I knew I needed to escape. I decided joining the military was the last-ditch effort to support my unborn child.

I went to one of those recruiting stations in a strip mall in my hometown—the ones that have all the military branches next to each other. I decided on the Navy because the Air Force and Army both denied me, and the Navy was the next door down. We had no access to the ocean in my home state, and here I am, about to prepare for a life at sea. Talk about a culture shock! As a child, I never passed sailors going about their way as I went to the store with my mother. Nor were there were ships to look at in the landlocked state where I lived. So, I had no mental images to relate to or draw from to help me adjust. I did not even know what the uniforms looked like; I just remember everything being so foreign to me.

But I progressed very well and made it through with a little trick I learned back home and cultivated on the football field as a student athlete. I turned my fear into aggression and the Navy loved it. The Navy loved it so much I was promoted to the leader of my recruit division. The military needs aggressive individuals, ones that will make decisions swiftly and carry them out forcefully. Aggressive people do not waiver, they do not take "no" for an answer, nor do they accept mediocrity. Aggressive people are confrontational; in fact, they often enjoy the confrontation because it poses another opportunity to defeat the odds, to prove the naysayer wrong, or to enforce their will. Unfortunately, aggressive people also tend to be prideful, arrogant, and narcissistic, because an aggressive attitude is more often rewarded as a sailor and is a sought-after trait for any U.S. Marine.

I noticed that in many professions, from firefighters to police officers to lawyers or business tycoons, aggressive people tend to become successful. However, the rewards and the accolades received for this aggression also comes with a price. It is a price that is often paid for with your emotion; it is bought with your compassion for others; and it is funded with your personal energy. It works well for a while, but as you try to juggle that energy and passion between your family and your job, the one that you perceive demands the most from you, is the one that tends to get all your attention and time. Listen to me: I did not say the one you love the most, but the one that you sense demands the most. I loved my family more than the Navy, but the Navy, through my skewed perception, demanded more from me. Let me explain.

> ...the rewards and the accolades received for this aggression also comes with a price. It is a price that is often paid for with your emotion; it is bought with your compassion for others; and it is funded with your personal energy.

In a normal job, if you do not like your boss, your job, or the people you work with, you can simply make the decision to clock out and not come back. This may sound extreme, but at least you can search for a new job, submit some applications, prepare your resume, and go on interviews. Afterwards, you can submit your two-week notice to your current job and then simply go work at a new place that gives you more enjoyment.

This is not the case in the military. Aside from my commitment to defend my country, as a provider to my family, I simply cannot decide I am not going to go on deployment. No good and upright person can make that decision, at least I could not. So, I must succumb to the demands of Uncle Sam who demanded that I leave for seven- and eight-month deployments. There is no punching the time clock in the military; it demanded that I be on call twenty-four hours a day, seven days a week. In times away, it demanded my full and undivided attention, so any thoughts given to loved ones back home could prove disastrous to myself or those around me. We work in extremely dangerous and hazardous environments, so if a sailor or soldier is not focused on what they are doing, if they are mentally distracted with what is happening back home, it could prove fatal. It demanded that I become cold and emotionally calloused to survive.

Emergency Room doctors that are distracted lose patients; SWAT team members that are mentally preoccupied get shot. While I do not pretend to understand these professions completely, I do understand that working under extreme pressure and stress requires some level of emotional callousness to get the job done successfully. As for the military, regardless of the branch of service, going to war and being on deployment changes a person. Training, drilling, and standing watch changes a person. Therefore, a mindset must be developed to perform, to be successful, and to stay alive. It is a military mindset.

Military Mindset

A MILITARY MINDSET IS regimented. It requires strict diligence, discipline that does not except failure or excuses, and a mentality that does not embrace fear or compassion. Furthermore, the military is an environment that tends to breed anger and aggression. Given the unique circumstances, some form of emotional detachment will usually take place for the service member to get through a naval deployment or to make it through a year in Iraq or Afghanistan. I am willing to bet that through the first fifteen years of my marriage, I missed over half of our wedding anniversaries, half of the birthdays, and half of the Christmas holidays. I learned early how to detach in order to get by.

My incorrect assumption that the Navy demanded more from me than my family led me to give that positive energy and passion that God intended for me to use for my family to the military mission at hand. The energy and passion your spouse needs, you can no longer provide because you no longer have it. There is no more emotion for your loved ones, no compassion for a hurting wife, and no more energy to work towards reconciliation because the successful completion of duties requires such an intense level of diligence.

I believe life in general does this to all of us, but it seems to happen quicker for many service members. For example, a man must feel respected and appreciated to feel important. If his wife, who is accustomed to leading a division or a company, becomes extremely critical and bossy, the husband may begin to shut down as he may begin to view his wife's actions as disrespectful. His wife, on the other hand, is just taking charge to "accomplish the mission." Also, consider the wife who needs some extra love and attention because her day with the children was a bit overwhelming and is now seeking security in her husband's compassion for her feelings. She may be met with insensitivity to her situation because the husband does not understand why she cannot deal with her problems the same way he does—by brushing it aside and moving on to the next task.

Both examples have one thing in common: the service member has not taken off their uniform, well at least not mentally. I understand to some degree this dynamic is true for most married couples, but if not careful, military members will begin to deal with all conflict the same way they learn to handle conflict on the ship or in the field—swiftly or aggressively, and sometimes both.

Naval sailors like other service members do not have time to dwell on emotions because we are not required to follow emotions; we are required to follow orders. We are expected to give commands for the good of all those involved—not to dispatch orders based solely on how we feel. So, while the husband hears the concern of his wife, he may be more inclined to try to move her past her feelings so she can "continue the mission" because she may still need to go to work or Johnny needs a bath or Suzy needs to finish homework. This is the military mindset I am talking about. Service members are expected to perform every task perfectly and under the extremely critical eye of their superior. Everything can always be done better; deficiencies are not acceptable, and any deficiencies found should be corrected with the utmost urgency. This creates an extremely critical and demanding attitude.

A man married to a military spouse could view this behavior as disrespect. However, the wife looks past giving him appreciation for what he has done and sees a deficiency that needs to be fixed so everything operates more efficiently. Efficiency eliminates downtime, rework, and wasted man hours. This is bred into every service member through even the most minute problems. A sailor, soldier, or airman that will pay close attention to the details of their uniform, equipment, or operations, will most likely do so with the lives of those around them. A questioning attitude and no leniency for discrepancy works well for combat purposes, but it is damaging to personal relationships.

It is impossible to bond with a person when you cannot communicate with them. It is hard to get close to someone who shows little excitement about things that the normal person would be ecstatic about. I found that for me it was hard to let myself get too attached, too involved, or overly excited. It is a mental chess game trying to decide what you will and will not allow yourself to emotionally engage in because in the back of your mind you know the inevitable awaits—separation. God understands this dynamic and knows a military mindset is damaging to relationships. It is also why He instituted there be no separation

> It is impossible to bond with a person when you cannot communicate with them.

between the husband and wife early on in their marriage. God knew they would need that time to grow together emotionally and spiritually so they will be strong enough to endure what their future holds.

Intimacy is the Key

THE REASON GOD SAID for the man to make his wife happy and not the other way around is because the woman holds the key to unlocking the intimate side of a man. When a man is intimate with his wife, she is happy, and it will be easy for her to give him what he needs which is respect and encouragement. Please understand intimacy is not sex; intimacy is communication on a natural and spiritual level.

This requires a man to listen to her concerns, to address her emotions without frustration, and to inspire her through their future dreams together. He must be able to speak to her heart in a way that makes her smile when she remembers what he said, make her miss him when he is away, and brings peace and security to her spirit. It is passionately expressing his wants and sharing his deepest fears. True intimacy is being naked before each other spiritually, where nothing is hidden, and no one is ashamed. Some have no problem being physically naked in front of others, but you could not pry their emotional nakedness open with a crowbar.

When a man can be intimate with his wife, it makes it easier for her to love him. It will be easy for her to nurture, support, trust, and encourage him when intimacy is present. She wants to be romanced, to feel safe, and to be rescued. She wants him to be her hero, not a lifeless robot. All women want this, even the military woman. God knows this must take place before the man goes off to war, because when he changes – and

> "When a man can be intimate with his wife, it makes it easier for her to love him. It will be easy for her to nurture, support, trust, and encourage him when intimacy is present.

he will – the only link back to what he used to be, will be through his wife. She reminds him of who he really is—the person who opened up and allowed her to see the real him. This is the inner man—the part of him that he so desperately wants to be but drastically contradicts who he is in his work

"

Please understand intimacy is not sex; intimacy is communication on a natural and spiritual level.

"

environment. This intimate connection is essential to creating and maintaining an emotional bond to one another.

This also holds true for female service members. Women in the military tend to be tough nuts to crack. They already feel as if they must work twice as hard to be respected amongst their male counterparts and are forced to constantly contend with feeling to prove their worth. This is true for many male-dominated professions: female firefighters are

> When she can encourage him, there is nothing her man will not do for her.

often viewed as a liability, a weak link, and unfortunately, the same is said about female police officers. The constant self-assessing, the continuous attempts to overachieve, always trying to substantiate their value creates a tough outer shell, an impenetrable force field to their hearts.

This can lead to them being extra critical and having little compassion for a husband who finds himself struggling at work or in life. She often will compare the way he handles problems with the way her peers do and conclude that he is inadequate. He may not be inadequate; he may just need a woman who is able to encourage him through the rough times and appreciate him for what he already does. Every woman must be able to find and embrace the positive things her husband does well, especially women of service. When she can encourage him, there is nothing her man will not do for her. This emotional connection is only developed when couples learn to speak to each other through intimate communication.

The inability to develop a connection, combined with the military mindset will create a divide between husband and wife leaving both parties unfulfilled. I remember I had no ability to communicate in a loving manner. Every conversation came off as harsh, sarcastic, and condescending. I did not engage in meaningful conversation much because I did not know how. Everything outside of what I did in the military became foreign to me, and I eventually came to the point where my wife and I no longer could relate.

This is a dark place many military families find themselves in: they struggle to come up with words like searching for socks to put on in the dark. Eventually, you just grab something to put on whether they match or not, and you begin to do the same thing in your relationships. Just throwing words out that do not match but you do so for the sake of saying something. It makes conversing awkward, unpleasant, and seem like a chore. I eventually begin to distance myself from my family because it felt increasingly like I no longer had a place with them. I fit more comfortably with the sailors I was around everyday than I did my own family, and it began to become evident in my expressions.

Application:
Talk: Discuss with your spouse how important he or she is to you and the difficulties of juggling the demands of your job with the needs of your spouse. Ask how you can be more effective in meeting the needs they have and list them below. Then be intentional in doing them.

_____

_____

_____

_____

Spouse: Discuss with your service member your support for them in performing their duties but express the lack of emotion or active aggression you have noticed in their daily interactions. Ask how you can help or what they need to reduce their stress and relax when they come home and list them below.

_____

_____

_____

_____

Chapter 2
Leaving and Cleaving

"I saw her across the room, and I knew I had to have her.
We talked often, and after sharing an intense moment
of deep conversation, I went home and told my family, "I
found my wife."

SHORTLY AFTER WE MET, Ellasin and I were at a party. I engaged in a conversation with her unlike any I'd had with another woman. I felt free and safe sharing my innermost thoughts and concerns. I was amazed at how she made me feel like I was important, as if I were destined for greatness. I can be extremely critical of myself, and after college initially did not work out, I was in a severe slump and feeling as if I were going to end up like everyone else in my hometown. You know the town, or at least you know the one where everybody knows each other, no business is private, and nobody seems to ever leave. I did not want to accept living in a low-income apartment working in one of the restaurants or factories in town while struggling to make it each month. I had not held down a steady job since leaving school about six months prior, and I was tired of being a regular at the temp agency.

Ellasin had a way of making me believe there was more in me than that. She talked to me as if she could see things in me I simply could not see, and she constantly challenged me to be better than hanging out smoking and drinking all night. I loved that about her; we had one of those type of conversations at the party that night, and I could not shake her from my mind. I knew at that moment we were going to be together no matter what I had to do to make it so. What I was feeling at that moment was a heavenly feeling, as if I were floating. It seemed as if she had a bright halo or glow around her face when I looked at her, and I found my thoughts becoming consumed with what she said, how she looked, and the way she walked. I thought this feeling would be enough to power us through marriage, through any problems, and would last forever. What I did not understand was that this was a period of infatuation; it was temporary and would not last.

When the heavens opened that night and light rained down upon her figure, I knew I was smitten, but did not realize that years later we would be in a brutal fight to save our marriage. When two people come together under the union of marriage, a process must take place to blend their lifestyles together. This is a natural process that all couples experience when learning to live together. It requires a lot of work for two people from diverse backgrounds with different experiences, beliefs, and possibly distinct cultures to learn and to live together in a healthy manner. Jesus alludes to this process when asked if it is lawful for a man to divorce his wife for any reason he chooses.

> "And he answered and said unto them, Have ye not read, that he which made them at the beginning made them male and female, And said, For this cause shall a man leave father and mother, and shall cleave to his wife: and they twain shall be one flesh?" —Matthew 19:4-5

Jesus's response suggests that to divorce the woman would mean for the man to separate a piece of himself, for they are one flesh. The relationship between husband and wife is to be closer than that of a child and his or her parents—no longer two separate individuals but one. Nobody hates his own flesh; nobody cuts off his own hand nor should he cut off his wife with whom he is now considered one flesh.

Divorce is now too often the quick fix for couples who think they are incompatible or somehow fell out of compatibility because they do not understand that marriage means to be united as one. However, the act of becoming one is a process Jesus refers to as "leaving and cleaving." This process of leaving and cleaving takes a much longer amount of time than you may lead yourself to believe. The process of *leaving* means to leave old ways, traditions, securities, and everything that is normal. It takes time to leave what is familiar, comfortable, and what feels secure. You may have lived with your parents for twenty years, and in that time, you have developed a cognitive process based on the beliefs of your family, the ethical values you were raised with, as well as how to communicate with one another. Therefore, it is hard to just leave twenty years of a psychological mindset you have developed.

To *cleave* to someone in the simplest terms means to be joined with them. It is to cleave to ideas you may not agree with, to customs that are foreign to you, or to someone that you may still have some insecurities about. Leaving and cleaving takes time, and the transition is often smoothed over by parents or in-laws who have experienced the rigors of life and the difficulties of marriage. Their sage advice can be utilized as a guide to navigate through the early years of marriage, until through much compromise, the couple is able to develop their own set of standards for their new family.

This makes the process of leaving and cleaving much harder for those who do not have the example of a strong marriage to draw from, friends, or loved ones who are actively engaged in a successful marriage. If you are someone who did not come from a family full of resources to pull from, do not be discouraged. Although this may be beneficial, it does not guarantee a successful marriage. You do! Regardless of your upbringing, military marriages face an uphill battle from the onset that most couples never have to endure.

Military Marriage

WHEN YOU MARRY A service man or woman, you also marry their obligation to their country. These demands will no doubt require you to relocate away from family who should be able to encourage and strengthen the marriage in its infancy stage. Imagine being a spouse from another country and having to relocate to the United States with no access to your family. When forced to leave what is most familiar, people tend to hold on to what they cherish for dear life. If a person is unwilling to give up something, there can never be a compromise. Many military spouses are uprooted from everything familiar, removed from the security of family, placed in unfamiliar or foreign lands, and surrounded by strangers. Even if both people were born stateside, at some point if the service member stays in, relocation is almost always guaranteed.

Therefore, most military couples often have trouble cleaving because they refuse to leave what is familiar. It may sound silly, but many service members and their spouses have argued over much of the same things every couple does such as why their family does not drink eggnog at Christmas or whether to celebrate Christmas at all. They may quarrel over the proper way to make potato salad or how to discipline the kids. Unfortunately, one thing that separates the military couple from all others is the requirement for many service members to deploy for long periods. Many military couples who have been married for almost ten years are still in the process of leaving and cleaving and they do not even know it. Leaving or letting go is often hard, but what makes it harder is if there is some insecurity in the one you are supposed to be cleaving to.

Scripturally speaking, to cleave means to join by gluing or adhering. The ingredients for the glue of a successful military marriage are communication and trust. Communication may seem easy in the beginning, but after a while, it takes work. When things are no longer new, we get accustomed to one another, and make the mistake of assuming that communication will "just happen." Communication must be initiated and entered into freely without fear or any inhibitions. You must be ready for sensitive subjects that bring up bad memories or lead to disagreements that become "touchy" topics. When touchy topics do not get discussed, they become roadblocks to communication making it feel awkward and unnatural. This is because there is no trust in that area. It is possible to trust a person in some areas but not trust them in others.

> The ingredients for the glue of a successful military marriage are communication and trust.

A lack of trust is like corrosion to a ship. You can walk by it every day, ignore it, or even paint over it and try to cover it up, but neither action addresses the root issue. If a lack of trust is not properly dealt with, it will be like that corrosion on the ship and slowly start rotting away the

"

A lack of trust is like corrosion to a ship. You can walk by it every day, ignore it, or even paint over it and try to cover it up, but neither action addresses the root issue.

"

relationship from the inside out. Communication becomes even more difficult when there are breaks in continuity. The husband of a successful businesswoman may deal with his wife leaving for business trips for a week or two. The husband of an Army wife who is a 12B (Combat Engineer) must live with her deployments being yearlong. Navy spouses see loved ones sail away for seven- to eight-month deployments. Both instances make consistent, meaningful communication null and void.

Some couples deal with unimaginable trust issues: some wives may question whether their husband really went to the gym for an hour when he left home. But the wife of a service member has no real idea what her husband is doing when he is in another country. The United States Military provides an excellent service to the people of our country, but it hinders the ability of its service members and their spouses to cleave or be adhered to one another. These stressors create a great challenge for all, but if unskilled at working together in a military marriage, it can prove disastrous. In fact, it does not matter what your profession is if you and your spouse are unskilled in working through your problems together, it will lead to problems in your relationship.

Fixing Communication

CONVERSATION MUST BE INITIATED. It may feel uncomfortable now, but if you take your mind back to the beginning of your relationship, you longed to talk to each other all the time. When catching yourselves falling asleep on the phone, one would say, "Okay, hang up," and this would be met with the reply, "No, you hang up." Then one would finally say, "Okay, on the count of three, we will hang up at the same time." One way to regain that closeness is to initiate a conversation about anything other than what you did at work.

As a Navy man, I found it hard to talk about anything except what I did at work, the ships underway schedule, or my gripes and complaints about having to rebuild equipment and personnel problems. While these things are not bad to discuss, when you have a job that removes you from the home often and for extended periods of time, your spouse may begin to see your occupation as an enemy to your marriage. My wonderful wife can tell you about many things that have to do with my job because while she was starving for attention, she still endured listening to my rambling about pumps, fuel tanks, and valves, to try to feel some sense of connection. She was willing to entertain conversations about something she was not interested in just to have conversation with me.

Unfortunately, this did not bring us closer together; it led to her feeling as if she no longer knew me outside of the Navy. The spouses of many first responders feel the same way; most people are not interested in hearing about accidents and injured people, but many endure it for the sake of conversation. As a service member, it is important for

> ...when you have a job that removes you from the home often and for extended periods of time, your spouse may begin to see your occupation as an enemy to your marriage.

you to initiate meaningful conversations about things other than your career. Talk about topics that interest you both; listen carefully and respond to the details being given. This shows you are interested in what your spouse has to say. It is easier to engage in healthy conversation that builds intimacy

You possess within yourself the ability to escalate an argument or to defuse one.... Words are like bullets; once you send them down range, you can never get them back.

if the responses are longer than one or two words. Positive communication will go a long way in advancing the cleaving process.

If the most critical component of communication is what is said, the second most critical component is how you respond to what is said. Tone and non-verbals are more important than your words. It's also important to be considerate of your words and self-aware of your facial expressions and body language. Remember, regardless of what may be said to you or about you, the truth is you have a choice in how you react to it. You possess within yourself the ability to escalate an argument or to defuse one. Therefore, when conversation becomes heated, do not use words that will hurt your loved one. Words are like bullets; once you send them down range, you can never get them back.

Rebuilding Trust

TRUST IS ONE OF those things that can be hard to earn, and once broken, it is almost impossible to restore. In fact, the only thing that will rebuild trust is time and consistency. Trust issues may come because of your spouse's past relationship experiences or something you have done to break their trust in you. Either way, consistency in your actions will go a long way to bring security to the situation. Those problems tend to be magnified in the wake of a service member's times of separation. It is hard to develop consistency if one is not around to work on it. However, being consistent and intentional with the time you have will go a long way in rebuilding trust.

Time is precious; it is something we have no control of. Once it is gone, that moment cannot be reclaimed. Since we all are limited to the amount of time we will have in this life, time is critical and valued above all things. A person who is terminally ill cares nothing for money; it is my experience that their wish is to have more time. More time to spend with family, more time to make amends, and more time to repair relationships. I encourage you to dedicate time to your spouse. The U.S. Army, Navy, Air Force, and Marines all get their time from you, but when the decision to spend time with your spouse is yours, it makes them feel desired and appreciated. The critical piece about time is that for it to be effective, it does not always have to be the amount of time spent, but what you do with the time that you have.

Application:
Leaving and cleaving requires compromise and work. Understand the differences and potential areas of disagreement within your relationship. Do not be afraid to talk about them and work together to reach a solution. Talking about the "touchy subjects" builds intimacy and security.

Fixing Trust takes time and consistency. Consistency is developed when your actions mirror your words.
- If you wronged your spouse, do not focus on how long it may take them to "get over" it; instead, focus on being consistent.
- Acts of kindness and active listening will enhance your marriage and help with the healing process.
- Remember quality trumps quantity. Fight to resolve; don't fight to hurt!

Chapter 3
All Men are Cowards

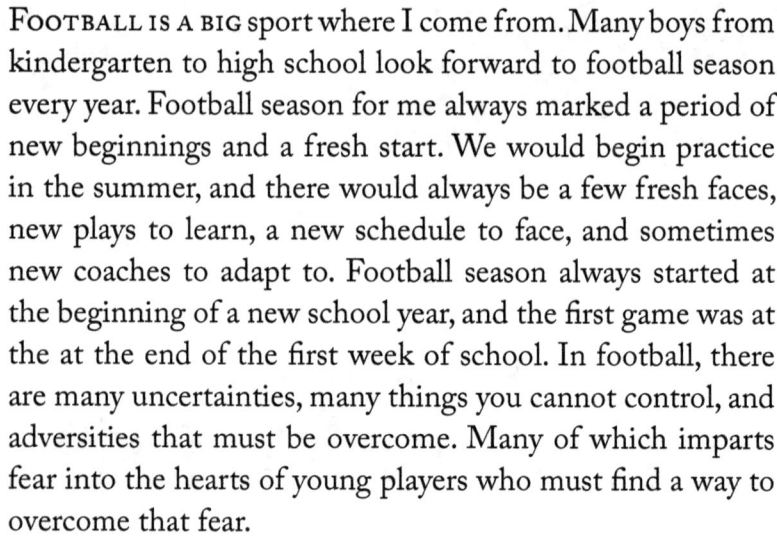

FOOTBALL IS A BIG sport where I come from. Many boys from kindergarten to high school look forward to football season every year. Football season for me always marked a period of new beginnings and a fresh start. We would begin practice in the summer, and there would always be a few fresh faces, new plays to learn, a new schedule to face, and sometimes new coaches to adapt to. Football season always started at the beginning of a new school year, and the first game was at the at the end of the first week of school. In football, there are many uncertainties, many things you cannot control, and adversities that must be overcome. Many of which imparts fear into the hearts of young players who must find a way to overcome that fear.

However, one thing is certain—it is not overcome on the football field by having a friendly chat with your opponent. In fact, if the other team senses you are afraid, they will have their way with you the entire game. So, we learn to hide our fear by being aggressive, attacking the enemy head on, and not backing down even though we may be afraid. Somewhere along their lives, most men have been introduced to the idea that displaying any emotion other than anger is a sign of weakness. Even godly men and men of great moral character are familiar with the notion that a man is to be the image and symbol of strength and power; and that in the attempt to confirm that image in their own lives, there can be no room for any signs of weakness.

I can remember when I was about four or five-years-old, I would love the fact that my dad would wrestle with me because he was showing me attention and spending time with me. I also hated it because my father was very rough in his attempt to "toughen me up." One day as we were engaged in our loving version of household combat, I fell, bumped my head on the floor, and immediately began to cry. My father who had been laughing and playing with me instantly displayed a cold, stern look on his face and said, "You're okay, stand up and wipe your face. Boys don't cry."

I never learned that it was acceptable to not be okay, to hurt, or to feel scared and insecure. I, too, am guilty of making such statements to my son during our times of play fighting. Not that I am totally against expressing the importance of being strong when in pain or in times of fear, but I wanted my son to be manly and tough, and I feel most men want that for their sons too. Nevertheless, I now know it is important to acknowledge the fear and recognize the pain. While you may press through those feelings and move on, it is extremely vital that you at least acknowledge they are there.

To be self-aware of personal weaknesses and fears will bring us face to face with our own limitations. Understanding these limitations will afford us the opportunity to be intentional about seeking help. This is a process that is not naturally learned by men, and very few know to teach it to their sons. I have come to realize when we teach our sons not to identify or address their emotions, we are teaching them to be cowards. Let me explain. Most men are not good at expressing how they feel. Not only that, but men often have a tough time even identifying how they feel to begin with. From the time we are little boys, we are bombarded with peer pressure to act like men, yet we have no idea how to do that. Through our experience, the best example of being a man is acting tough and showing little to no emotion except for anger.

To feel an emotion such as fear is frightening because we were taught to suppress those feelings and be strong like a man. Therefore, it is frightful to express or identify a feeling that may make us feel weak or inadequate. Instead of facing our feelings and learning to identify what they are, we teach our sons to steer clear of their feelings, leaving them to be cowards with their emotions just like us. Unfortunately, when this happens, we tend to stumble through life and our relationships with a couple of emotions we are familiar with and then disregard the others.

There is no place this is more evident than in the United States Military. There are many "sea stories" from old sailors based upon what they have done and where they have been. They are often extravagant tales of past port visits and fables of legendary sailors who were superheroes within their field of work. There is one such story that sticks in my mind, and it is relevant to this topic. It is of a young sailor who is met by his supervisor on his first duty station onboard a ship. In getting him settled into the division, the supervisor proceeds to tell him that he is a "hands on" supervisor. The new recruit thinks he works with the crew to accomplish the job and ensure it is done correctly. However, the supervisor's description of being hands-on was, "If I don't get what I want, I put my 'hands on' people." When asked why he was like that, the response is something to the effect of, "I am tired of getting in trouble because people are not doing what they are supposed to do." He was held accountable for the actions of his subordinates, and if they were deficient in job completion, he would get in trouble for it. He met his fear of being reprimanded by his superiors with anger and aggression towards those he oversaw.

Our job is a tough one that does not allow a lot of room for feelings or emotional expression. To do so would allow others to perceive weakness, suggesting that you could not be counted on to get the job done. For example, in the exchange of gunfire during a firefight, they would expect you to run instead of covering their six (back). Fear adds no benefit when called upon to save sailors who are trapped in a burning compartment or space on a ship. Because of the unique expectations of military service members, married military men often have a challenging time expressing any emotion towards their family other than anger or frustration. Any disagreement with a spouse, any insecure feelings within the marriage, or any fears the man may have are expressed through anger. While we think we are displaying strength, we are really showing that we are cowards afraid to admit our fears, weaknesses, or express our feelings of love. Therefore, emotionally speaking, men are cowards.

How Men Love

TIMES ARE CHANGING, WOMEN are more independent in this country and in the world than ever before. They own their own businesses, purchase their own houses, and buy their own cars. Even in this modern age of gender equality, many men still feel they must fulfill the role as protector and provider. Ladies, it's not that we think you are incapable; in fact, your capabilities may be what attracts us to you, but we still feel the urge to prove our devotion by what we do, instead of what we say. Men tend to be goal driven and the achievement of that goal adds value to our self-worth. It tells us we did an excellent job, that we are advancing, and are destined to be successful. Success ensures more money which in turn guarantees the family will be taken care of. Therefore, we are driven to accomplish the task no matter what, reach the check point no matter what, and achieve mission success no matter what. Truthfully, emotional connection has truly little to do with a goal-driven individual.

Quite frankly, emotions can become a hindrance; it will prevent you from pushing towards the goal because of how you may feel. Goal-driven people suppress their emotions to accomplish the task. This will lead people to neglect their emotional responsibilities to their family and spend more time at the office. Civilians do this out of a desire to be the best; the United States military does this out of necessity to survive. Many jobs we do are dangerous, high-risk evolutions with high probability for loss of life or valuable assets. And daily we suppress our fear, our loneliness, and even our love, because any emotional distraction could have catastrophic consequences. There is no doubt that a man feels love; however, how to express that love is often far from us. We express our love by what we do—we had rather buy expensive gifts than to try to find the words to articulate our affection. We are great at doing, but we suck at talking. Where military men get into trouble is by using empty words to express how we feel; it may not be how we truly feel, but we know you will like the way it sounds. We profess the love part because we remember what to say, but our hearts often are trying to connect with the words coming out of our mouths. We know what to say because you taught us; we know how to tell women what they want to hear when dating them because of the responses we receive, but many times we have no emotional connection to our words. It puts a strain on the communication and creates a roadblock to intimacy. This is the same way God felt about the people in Isaiah 29:13.

"Wherefore the Lord said, Forasmuch as this people draw near me with their mouth, and with their lips do honour me, but have removed their heart far from me..." —Isaiah 29:13

When we replay Israel's actions within our own relationships, it always leads to no intimacy with our spouses. To be intimate with somebody has less to do with sexuality and more to do with an emotional and spiritual connection. To be intimate with somebody requires you to expose your inner most feelings, to share the things that you may be afraid of, expressing your insecurities, and proclaiming the way you may feel about them. God's problem is the same many women have with their husbands. Husbands know their wives well enough to say and ask the right things on que, but without any heartfelt emotion attached. If she looks sad, he may ask, "What's wrong, babe," although he may not care to hear the intense details of her struggle, but he asks because he feels it is the right thing to do. Or in the morning before they go their separate ways—either to the base or the office—the kiss goodbye accompanied by the lifeless words "I love you," could easily have been substituted with a handshake. Unfortunately, many men travel the road of the coward, telling their spouses what they want to hear but concealing what's really in their heart.

God was upset because the people knew the right words to say and the right things to do, but their hearts were

To be intimate with somebody requires you to expose your inner most feelings, to share the things that you may be afraid of, expressing your insecurities, and proclaiming the way you may feel about them.

not in the things they were saying and doing. It had become regular, routine, and ritualistic. God hates ritualistic love, and so does your wife. Trust me, she knows if your heart is not in what you say. But many attention-starved wives will endure the ritualistic love expressions to desperately cling to some sense of a connection with their husbands.

Ritualistic Love

WHEN THE EXPRESSION OF your love becomes ritualistic, it will lead an individual to become selfish in their actions—only expressing heartfelt emotion when they want something. Israel made this mistake with God. The Israelites became just like a man trying to pick up a woman at the club; they were trying to tell God what they thought He wanted to hear to get what they wanted. Men do this, boyfriends do this, even husbands do this. Our fathers, friends, and society has taught us to treat woman like a goal to achieve, a point to score, or a game to win. We equate women with prey, and men are the hunters. Hunters often use decoys that confuse or draw their prey in to make them feel safe. Men lie; it's what we do. It's part of our arsenal of weapons to capture our prey. We have grown up using words as decoys to trick women into believing us long enough to get what we want.

In the beginning of a relationship, men really mean what they say. The reason a lie and the truth sound so much alike is because the lie is what men sincerely want to feel without allowing themselves the liberty of truly indulging in those feelings. A woman they eventually fall for gets the truth because he feels safe enough to engage in conversation with her and share his feelings. Sadly, in many marriages, the security husbands feel during early conversations at the beginning of the relationship is replaced with hesitancy because they perceive their wives only complain about what they are not doing right. They no longer feel safe, so they no longer engage in the "deep conversations" like when the love was new. A man who does not feel safe will not talk. You will get some of the same words you used to hear, but they have removed their emotions from them to protect their own heart. A person who withholds conversation because of their personal feelings is operating in selfishness, and they rob their partner of the one thing they need to stay connected—intimacy.

Selfishness leads to ritualistic love, only saying certain things to keep the peace, to get their way, or to avoid conflict. Because some men have not learned to express their inner most feelings with their spouse, they feel insecure and afraid to share their real feelings. Revealing their vulnerability makes men feel weak, therefore, they are cowards when it comes to expressing the heart. They had no practice growing

> A man who does not feel safe will not talk.

up, and they were not taught how to identify what they feel. What if I told you that identifying your emotions and expressing how you feel does not make you weak at all; it makes you human. Away with this thinking that men are emotionless robots who have no sympathy or compassion. Men have very real emotions, real care, and real compassion; the problem is identifying those emotions and expressing our concerns through verbal communication in healthy ways. I encourage you to not be afraid of the emotions, but you must understand how your emotions affect your interactions within your relationships. Don't be a coward. Some men do not talk as much as women, but men do need to learn how to talk to the woman they love.

> Men have very real emotions, real care, and real compassion; the problem is identifying those emotions and expressing our concerns through verbal communication in healthy ways.

Application

Men: For one day (because a week may be a bit much), write down several emotions you experience other than anger and how they make you feel. Here is the challenge: now share that with your wife.

Women: If he shares what he feels no matter how silly it may sound to you, do not laugh, do not get upset or angry, and do not criticize. If you want him to talk, create a safe environment for him to do so.

Chapter 4
Red Tags

"Wherefore, my beloved brethren, let every man be swift
to hear, slow to speak, slow to wrath: For the wrath of
man worketh not the righteousness of God"
—James 1:19-20

It is a well-known fact in the Navy, when speaking
regarding maintenance procedures, that the Navy's safety
regulations were "written in blood." As the Navy advanced
from wooden ships powered by wind and sails to a fleet of
steel behemoths powered by engines and electrical
components, there became an instant need to maintain
these mechanical parts. Machinery, engines, and electrical
equipment possess significant levels of risk to bodily harm.
The developers and manufacturers understood and advised of
such risks during manufacturing, but there were unforeseen
risks that came about once that equipment was placed
onboard ship, put into service, and sent out to sea. As a result,
many people were severely injured or killed operating or
maintaining equipment aboard ship. This led to the writing
of regulations and safety rules as the Navy experienced these
growing pains of evolution and led to the above-mentioned
phrase that "Naval safety regulations were written in blood."

It's important to stop right here and mention that
marriage, as it is designed, is wonderful and the most

rewarding experience you can ever have. But just like the equipment on a ship, when we get in the real world and must navigate through marriage, we run into problems. We find out very quickly that maintaining our marriage in good working condition requires us to constantly work on it. Simply put, we must do maintenance. Every branch of the military must do maintenance on their equipment, and there are safety measures put in place to prevent bodily harm or death when doing so. In the Navy, that process requires shipboard sailors to hang Red Tags.

Hanging a red tag on a piece of equipment is done to prevent a casualty by significantly minimizing the possibility of severe bodily harm or death. That is because there is a system in place that, when properly utilized, will prevent operating equipment that is damaged or marked for maintenance. Aside from the proper verification of equipment and maintenance procedures, the Red Tag that is to be hung has the word DANGER written on it to prevent sailors from operating the equipment while performing scheduled maintenance or repair. Like equipment in need of maintenance or repair, some emotions will require you hanging a Red Tag on them to save your most valued relationships.

Emotional Maintenance

EMOTIONS ARE A PSYCHOLOGICAL state influenced by thoughts, feelings, behaviors and various likes and dislikes. They are feelings that can be inadvertently triggered by a person, place, or thing causing the individual to produce rational or irrational thoughts. Emotion or feeling creates a thought in the mind that motivates the individual to react or respond in a certain way resulting in a behavior that aligns with the emotion. For example, if a friend gives you a card or flowers, it can trigger an emotion of feeling happy. This produces a thought of appreciation, so your behavior is displayed by a big, beautiful smile as you say, "Thank you" or "This is so nice."

It can be said that emotions are the fuel behind our cognition which affects our feelings and our behavior. In this same scenario, the card or flowers are given, but this time they trigger a thought of sadness because the last time you received flowers was from a previous partner right before they dissolved the relationship. The thoughts of what they did trigger you to react in a negative manner and throw the flowers away. The behavior does not match the event that took place, because irrational thinking created emotional distress.

Too many military members operate in the irrational thinking department of their minds, placing them in a constant state of emotional distress. When you are unable to identify what emotion you are feeling, it is often manifested through a surface emotion such as anger. I am the worst in this area; anytime I feel threatened, challenged, or afraid, I want to react in a display of over-the-top anger. I used to contribute this to the Navy—the fear of being wrong on an inspection or facing the consequences of discrepancies from a certification event would make me feel insufficient or uncertain. I did not act in fear though; instead, I used the surface-level emotion of anger to ensure I felt secure. I used it to take control of a situation I had no control over. I came to the realization that this did not come from the Navy; this came from an experience as a child that left me feeling like I always had to protect myself. So instead of being scared, frustrated, or nervous to protect the little boy in me, I would go into a rage and get angry. The military mindset I developed awarded me for this behavior because it got results, but it was damaging to myself and my family.

As I reflect on my own career, the anger I displayed would not allow my family to emotionally connect with me when I was at home. When my feelings were hurt, I got angry; when I was uncertain, I got angry; I even got angrier when I was angry. My thoughts became more irrational, and I began to feel as if I did not belong, and that my own family did not want me around. By not identifying these feelings, I continued to fester, and the more I festered, the more infuriated I became.

Resentment caused me to isolate; my communication was of few words if any at all. My actions did not cause my family to stray from me, they were pushing them away and causing a disconnect. At this point, you would think I became more self-aware and began to ask myself what I was doing wrong. No! I did what everyone who has ever relieved a previous supervisor has done. Instead of taking responsibility, I blamed my current problems on the guy that left. In other words, I justified my actions by selfishly projecting blame on everyone else without taking any responsibility for my own. This created a rift so large I truly felt like a visitor in my house. However, it was my beautiful wife's patience, grace, and willingness to collaborate with me that helped me identify the underlying emotion was a feeling of rejection not anger.

Emotions are like different pieces of equipment that when operated can dictate a person's mental state. We need to understand that some of our emotions may require us to place red tags on them when we find ourselves in certain situations. Using these emotions is like operating a piece of equipment that may be tagged for repair—that is because they are a DANGER to your marriage and can lead to devastating consequences if ignored. We all have a past, and the history of our lives is made up of both good and bad moments that work to shape us as individuals. However, because of some of these moments, there may be areas in your life that are damaged, and if not careful, operating in the wrong emotion during times those parts of you are exposed could lead to danger. For example: your trust has been damaged due to an extramarital affair or your spouse maxed out all the credit cards while you were on deployment. After heart felt apologies have been given and forgiveness has been rendered, you may want to put a Red Tag on your anger. It is not a crime to get angry; is a normal human emotion. But it is how we react when we are angry that creates problems.

Many times, what happens when a spouse gets angry is they will bring up how their husband or wife hurt them in the past, and it will have nothing to do with the current situation. Putting a Red Tag on your anger means you become aware and

> It is not a crime to get angry; is a normal human emotion. But it is how we react when we are angry that creates problems.

are conscious of what makes you angry, then avoid lashing out inappropriately when you feel yourself about to blow a gasket. It will prevent you from saying and doing things to intentionally damage or hurt your loved one.

Maybe you struggle with showing or expressing compassion—I know I did. A person who lacks compassion will blow off their spouse's feelings, concerns, or hurts leaving them feeling unvalidated as if what they are feeling doesn't matter or isn't real. People lacking compassion will often say things like, "You need to just get over it." Sorry friend, but you should hang a Red Tag on your lack of sympathy and understanding. Any emotion can become a toxic one if utilized in the wrong way. There are times when we face negative situations, but negativity should not be part of our personality. It breeds contention and bitterness for the spouse married to the one who always finds something wrong with everything that is done or has something negative to say in every conversation.

Have you ever been near someone that seems as if they just sit around and think up negative stuff to say? It is hard to be around people like that, and it is extremely arduous being married to one. There are many people in the military that are condescending and sarcastic individuals; it is a trait we use to deal with unwelcome news or harsh situations. We pump ourselves up to be greater than we are and make light of undesirable circumstances to motivate ourselves to finish the job.

Over the course of my career, I have seen sailors who have been critically injured. I remember one sailor who severely hurt their hand to which the Chief responded, "Thank God you have two; use your other hand." The problem with being in this environment

> There are times when we face negative situations, but negativity should not be part of our personality. It breeds contention and bitterness.

and listening to this kind of daily rhetoric is it becomes accepted as normal conversation. Once the condescending and sarcastic comments become normal to the service member, this same toxic communication eventually find its way in the conversations at home, with family, and with friends.

A wife dealing with a problem needing the compassion of her husband finds no help and security in the man she is married to if she receives a sarcastic comment because he is unable to find words to ease her mind. If you tend to be negative or have a negative outlook on life, put a Red Tag on it, and seek counseling to help you view things more clearly. There is nothing wrong with counseling; in fact, more of us should take advantage of it to improve our mental health. If you are truly able to talk to your spouse without expressing those hard emotions, it can heal present communication breaks and prevent future ones.

Maintenance Complete

When a Red Tag is removed, it indicates the work is complete, the maintenance or equipment repairs have been accomplished, and it is safe to operate normally. It should be noted that one person can never remove a Red Tag; it takes a maintenance supervisor to authorize the maintenance person to remove the tag. Removing the tag is a two-man process that requires the second person to verify the first person removed the right tags. It is the same process as hanging the tags as well; it is a fail-safe way to ensure the tags are not hung on the wrong equipment and to ensure the proper tags are removed when the maintenance is complete. This verification process eliminates any further damage to the equipment or potential harm to personnel.

Service members and first responders are exposed to traumatic events and unsettling scenery that led to unhealthy ways of processing those experiences. This is the reason for the widespread alcohol use through the military as members seek to self-medicate themselves. There are many ways in which to self-medicate after the moment, but it is what a person says to themselves and to others in the moment

"

There is nothing wrong with counseling; in fact, more of us should take advantage of it to improve our mental health.

"

that moves them through the situation. These situational comments should not be used in our interaction with family. These cold and calloused comments displayed to loved ones are a product of behavior developed by the service member's absurd thought process as they emotionally deal with their daily experiences.

Alas, we can become familiar with the behaviors we have and even view them as normal. You may think you have fixed or repaired certain areas of your life because you do not address them. However, emotional distress is our bodies' warning sign that there is a problem within us that needs to be addressed. I want to encourage you to talk with your spouse to see if you operate in emotions that lead to harmful, toxic, or dangerous behavior in the relationship. Learn to sit with and face yourself. Become more self-aware, acknowledge those thoughts that lead to toxic behaviors, and seek to identify where they come from. Try asking yourself this question, "Is this really how I feel, am I this angry, sad or disgusted, or is this a surface emotion to something else?" As I have stated before, there is nothing wrong with being angry; it is the behavior you exhibit when you are angry that determines much of the outcome. This can be said for other feelings and emotions as well.

Be Prepared to Hear the Truth

ONE OF THE HARDEST things a person must do is be real with themselves and take responsibility for their actions. When asking your spouse about the behaviors that become known when expressing different emotions, you must know that you may not like the answer you receive. That is because we do not like for our deficiencies to be exposed. It is the examination of these behaviors that connect us to the feelings and emotions that cause them. Once you can identify what triggers those emotions, you can be more intentional with how you act while in the heat of the moment. This is how you mentally put a Red Tag on that emotion or on that behavior. However, the key to understanding where these tags should be placed is in the honest communication with your spouse. These conversations are hard because we do not want to feel as if we are being criticized or if we have let our spouse down.

Military service members tend to make these conversations more difficult to have. It has been my experience through the countless inspections and certifications throughout the years, that we do not like discrepancies. Discrepancies in the life of a service member are frowned upon. Because we have such ambitious standards, our tolerance for discrepancies is incredibly low; when we are found with a discrepancy during a certification or inspection, we tend to take it personal. There is rarely an "Oh, we found these problems, so now you know what to look at" approach. It is more

> There is nothing wrong with counseling; in fact, more of us should take advantage of it to improve our mental health.

of a "We found these things wrong and that means you are all jacked up because it's not perfect" approach.

The conversations with your spouse, however uncomfortable they may be, must not be intended to determine whether you are incompetent or purposefully neglectful in how you interact in your marriage. They should, however, point out what is preventing you from being better than what you already are. Trust me, if you ask your wife, she will let you know if you have gotten better in certain areas, and by asking, it shows you value her opinion and are taking steps to be better for her. If you ask your husband if you are showing progress in how you interact with him during certain emotions, he will tell you, but it may sound more direct than you would like. However, the key in both situations is the spouse must feel secure in being able to bring it to your attention without you taking it the wrong way. This takes practice on both parts, but the more you work at it, the better you get. It has been twenty-four years, and I am still working at it, so you are not alone. Let us all continue to work at being better for the ones we love the most.

> The key ... is the spouse must feel secure in being able to bring it to your attention without you taking it the wrong way.

Application:

Everyone communicates, but is it healthy? To develop healthy communication, you must become self-aware of how you currently communicate with others and be willing to modify or change it if it is deemed unhealthy.

For one week I would like you to put the following into action:

- Be concise when you communicate. This will cut down on misunderstandings.
- Actively listen. This means do not listen to defend or give a rebuttal, but to understand. Then paraphrase or summarize what your partner said to ensure you have clarity.
- Remember you cannot control others; you are only responsible for your role in a relationship.

Chapter 5
Deployment

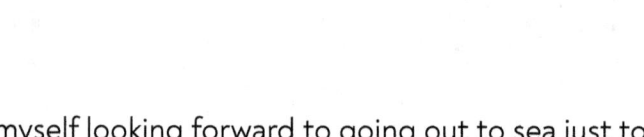

"I felt myself looking forward to going out to sea just to get away. The sailors understood me. I fit in there. This has now become my life because the Navy has become my priority."

THERE IS A WORD that carries with it a myriad of emotions when talked about. It is a word that is filled with pride and anxiety, devotion and selfishness, joy and sadness, anticipation and delay. The word is *deployment*. It is a word that means something a little different to every person you talk to. Some will say it is our devotion and duty to America; others will talk about how we have no business in foreign countries when we have problems at home. Given the "binnies" (benefits) like extra pay for hostile environments or tax-free money, there are those who look at it as an opportunity to make a lot of extra money. Others see it as a chance to finally get into the shape they want to be in. During deployment, other than the mission, there is not a whole lot to do besides going to the gym. Then there are those who will say it's the uncertainty of the mission that may make them fearful or the long working hours we are all accustomed to while deployed.

There is one common thread that you will find amongst every service member you ask about deployment, and almost all will respond with dreading the time spent away from loved ones. As I look over my twenty-four year career, I can honestly admit I have missed over half of our wedding anniversaries, birthdays, and holidays. Half! Think about that for minute. Along with those important days, what about the "first time" events that only happen once and call for a celebration. I have missed my baby girl pulling herself up to stand and cautiously taking her first steps. I missed taking my son to his first day of kindergarten, and my oldest daughter's modeling auditions. Others have missed the first tooth coming out, points scored in a football game, recitals, church solos, proms, graduations, and so on.

There are a lot of things the service member misses out on being forward deployed—things that can never be replaced, things that put a strain on relationships, adds to frustration, anger, and unforgiveness. Deployments are hard enough leaving the family on good terms; it is almost unbearable leaving when you are at odds with a loved one.

> Deployments are hard enough leaving the family on good terms; it is almost unbearable leaving when you are at odds with a loved one.

There are very few professions that can relate to being separated for extended periods of time, but I am not aware of any that separate their employees, intentionally placing them in life-threatening situations and under extreme circumstances. One thing I do know for certain, and it is the same thing every other service member knows, is that deployments are extremely hard on families. It does terrible things socially and psychologically to the family, and if you are not equipped to face the trials of a deployment and grow together, you will most certainly grow apart. The habits and behaviors we develop while apart can create long-term problems that lead to a toxic home environment, so much so that many refuse to endure the trauma and choose to terminate the marriage. But you can never terminate the children or the other family members involved.

It is an interesting dynamic because what is really being said is, "I refuse to continue to give up any more of my time." The spouse no longer sees the value in waiting for a husband or wife that never really comes home mentally or emotionally even after deployment is over. A service member may refuse to give up sweating, bleeding, and living a lonely life away from a family that does not appreciate the price that is being paid for their well-being. Either way, the truth is the time together lost because of separation is more than just time lost; it is the losing of one's life.

Time is critical; it is how we measure life. We all know we have a limited supply of time before we check out of here, but none of us know exactly how much. Since none of us will live forever in this body, we are all born dying. Therefore, we hate wasting our time on things that we do not want to be bothered with. Time cannot be stopped; it cannot be controlled. In moments you need it to move swiftly, it slows down, while other times you cherish seem to move too fast. Time dictates an enormous part of our lives—when we go to school, when we marry the love of our life, when we plan to have children, when we expect to have a wonderful job, when we hope to be promoted to that higher-paying position, and so much more. We are always in a hurry trying to stay ahead of time, yet always find ourselves behind it. The thought that time is always shorter than we would like it to be is always lurking in our mind. Some things can never be replaced, one of which is time.

Time Continues to Tick

WHEN I WAS A young sailor before the advent of social media and frequent Internet usage, an interesting phenomenon took place when on deployment. Time stood still. It was as if we were inside of an invisible barrier and the entire world continued to move forward, continued to grow, continued to produce new things, and we were somehow frozen in time for six to eight months. It is like living on a small parcel of land or floating around in a forgotten sea or undiscovered part of the ocean that time forgot about. When you have no connection to home except for a twenty-minute phone card every couple of weeks and snail mail (Yes, actual handwritten letters.), you would not believe how much changes while you are gone. When returning from deployment there are new cars, new movies, unique styles of clothes. Everywhere you look, you would see something new leading you to immediately ask, "When did that come out?" People die, relationships end, and you are left trying to find out at what point did things change. You question where you were and what you were doing to use as a reference point to try to draw some sort of a connection.

Service members often reference events that happen with where they were stationed or deployed at the time it happened. Most people say things like, "I remember this song; it came out in the summer of 2002." The service member says, "I remember this song, I was in Bahrain when it came out" or "I was onboard the USS John C. Stennis when that movie dropped" or "I was in Djibouti for her birthday." The service member is faced with the realization that while time may be standing still for them, it is quickly ticking away for the rest of the world. The military robs us of the most valuable asset to human life … time. Time cannot be controlled nor can it be increased. We cannot withhold the grains of sand that quickly slip through the hourglass. The best we can do is try to manage how we use what time we have left.

One of the greatest motivators for a service member is liberty. In the Navy, liberty is the term used to denote off duty hours, when you are off work. It means you are at liberty to leave the ship outside of work requirements. As a young sailor, I discovered an interesting phenomenon that swept through every branch of the military. I found out early that if you want a job accomplished, keeping sailors till 1600 will not get it done. Amazingly, promising them liberty as soon as the job is complete, will get an eight-hour day worth of work done in about forty-five minutes. That is because the sailor, just like the marine, the soldier, and the airman, all value their time.

Deployment is not a thief, but more like a robber when it comes to time. A thief will sneak in when you are away or while you are sleeping unaware and quietly steal all your valuables. A robber on the other hand is more desperate—they violently get right up in your face, hold you at gunpoint, and take from you what is most important. Because of the demands of government interests and national security, our military obligations viciously rob us of time with our family. As a result, it is imperative that remaining engaged with family back home be the single most important thing you do to maintain a healthy relationship with loved ones.

It requires the soldier, sailor, or marine to be intentional with their conversations and with their communications. Too

"

It is imperative that remaining engaged with family back home be the single most important thing you do to maintain a healthy relationship with loved ones.

"

many times emails being sent home to wives and husbands become vanilla and bland. "Hey Babe, how's it going? Not much here just the same old thing. I hope you have a good day. Love you." Something begins to happen that neither the service member nor the loved one back home realize at first. Even though emails are coming and going, communication –real meaningful communication –is breaking down because there is no quality time being spent together. There is no way to create more time, and the time you do have is spent away from the ones you love the most.

Become Intentional

THE SERVICE MEMBER IS asked to do amazing things: perform life threatening jobs on a regular basis and maintain the administrative side of the military to ensure everything runs according to regulations and instructions. Senior leaders counsel younger service members, issue corrective measures, and train personnel for their assigned tasks. We are asked to train, to fight, and even to die for our country, but we are

> We are asked to train, to fight, and even to die for our country, but we are rarely encouraged to plan dates for husbands or wives, go on vacations with our children, or taught how to cultivate personal relationships.

rarely encouraged to plan dates for husbands or wives, go on vacations with our children, or taught how to cultivate personal relationships.

In the Navy as well as the other branches of service, when dealing with a service member with marital issues, we have the luxury of sending them to see the Chaplain. But many Chaplains are tired, burnt out, and are dealing with their own issues with no one to talk to. They are overrun with young service members being sent to them to figure things out their supervisors still have not yet learned, such as how to deal with personal stress, how to cope with losing a grandparent whose funeral they cannot leave from deployment to attend, or how to successfully stay married and be in the military.

One thing that would help many marriages would be if the sailor, airman, marine, or soldier learned how to be intentional with their communication. It requires creativity and proper planning, but without it, the emotional connection becomes weaker until both parties become emotionally and spiritually withdrawn from one another. I can remember emailing my wife once a week, and I would say it was not because I needed it, but because she did. Truthfully, I needed it just as much, if not more than the family did. I did not perceive it then, but I needed to know they missed me, loved me, and that I mattered. Instead, my emails became another task I had to accomplish liken to completing another check on the maintenance boards. At times it felt quite daunting, somewhat equivalent to writing a term paper in college. It became increasingly hard to come up with the words to write, the words to express how I felt, or what I was experiencing, and it left me wondering why.

On deployment, time does not seem to advance, and things never change because of the routines we develop while we are deployed. Yes, there is a job to complete and a mission to accomplish, but honestly, the real reason we develop our personal routines is because it keeps us busy and takes up the hours in the day. It is too painful to count the days to returning home when you leave for deployment. There are too many and can be overwhelming. A routine keeps you focused on the next thing on your list for the day. If you want to see a service member lose their mind, change their routine. It throws them off and makes the days seem longer. The unexpectancy and the unknown gives you time to think, time to miss home, and therefore, time to feel sad. There are some military members who love routines because of what their mission requires; but for most of us, it is a means to pass time especially when there is no way to communicate. Life on deployment becomes so robotic, the service member loses any stimulation for new conversation of their own. For me, I was just responding to what my wife said and never really had any new input aside from the mission. In respect to the spouses left at home, both men and women alike often feel as though they are not thought about. As an example, it can leave a husband who is at home with the children feeling forgotten, as if his wife doesn't love him enough to talk about anything worthwhile. It sounds crazy, but honestly, there is nothing much to talk about. The scenery is always the same, the food is always terrible, and the hours are always long. You are always tired, and everybody is ready to go home. Try saying that a unique way every day for over 230 days. No matter how creative you are, it becomes a chore to talk about anything that has any real depth or meaning. So where does that leave your marriage?

After years of growing apart due to deployments, sea time, and duty days, my wife was able to help me understand the importance of sharing things together. One activity we came up with was to pick out a book together, and while I was at sea, read a chapter each week, make notes, and email them to each other. Then we would respond to each other's notes and send them back. We decided on a devotional book about marriage and did different bible studies together. Choosing one these examples or something similar will help keep you engaged in something new. It gives you a topic to focus on together and allows you both to express your thoughts on the matter. Doing things like this will allow you to keep open lines of communication and remain engaged while you are away. The key is to be intentional with your conversations, and this will give substance to your emails and phone calls.

Acts of kindness are another way to be intentional. This will show your spouse you are thinking about them even on the other side of the world. Before you leave for deployment, buy a bunch of greeting cards for the holidays, birthdays, anniversaries, and some just to say I love you. Send one each month letting her know how much she means to you. You could pre-order a new video game or arrange pizza to be delivered every two weeks for two months. You know he is stuck with the kids, at least let the man play the new football game and arrange for the kids to be fed. That way everybody is happy. For the service member, it is imperative you send flowers, write poems, or do whatever it takes to ensure family knows how you feel.

> Before you leave for deployment, buy a bunch of greeting cards for the holidays, birthdays, anniversaries, and some just to say I love you. Send one each month letting her know how much she means to you.

Unfortunately, the hard truth is the odds of us leaving for deployment and returning home alive are a lot slimmer than most civilians returning home from work at the end of the day. The best thing we can do is ensure that if we don't make it back, our husbands and wives have no question as to how much we love them. Time is ever fleeting from our grasps; be intentional with how you spend yours.

Application:
The next time you are underway, in the field, on deployment, or away from your spouse for the day, write an old-fashioned letter on real paper expressing how much you appreciate them, how important they are, and what they mean to you. Then mail it to them. If you will be home soon, you can give it to them face-to-face and watch their eyes light up as they read your words of affection.

Service Member:
- Write down three things that scare you the most about being away, then share it with your spouse.

_____

_____

_____

Spouse:
- Write down three things that scares you the most about your service member leaving for training or deployment.

_____

_____

_____

Reassure your husband or wife in these areas of concern so they feel more secure about the things they fear the most. The truth about your fears will cause you to become closer and will lead you to draw strength and security from each other. This is extremely important during times of separation.

Chapter 6
What About the Children?

"I hate you! I hate you because you leave me!"

"THESE KIDS ARE NOT your sailors!" Oh, how many times have I heard that one from my wife. I did not realize—and I am willing to bet many other service members do not either—that even though I was home in pajamas or jeans, I had still failed to take my uniform off. Mentally, I was still in that hard, callused, stay on the grind, always intense type of mode. So, I had a bad habit of speaking to my children like I did the sailors onboard my ship. Sailors, like many other soldiers, airmen, and marines, tend to talk *at* people instead of talking *to* them. It is more than likely a result of the nature of the job that requires orders to be given with the expectation that those orders are going to be carried out.

There is no room for debate when orders are given, and there should be no questioning or rebuttal unless the order given is an unlawful one. Every service member is accustomed to receiving and giving orders. It is so woven into the fabric of our being as military men and women that we do not realize the sharpness and direct tone in which orders are given. Because we understand orders are to be carried out expeditiously and without haste, we do not question the cold tone or the emotionless way the orders are given. We just do what we are told.

This is a dynamic that most of the world does not understand unless they have served in the military. Service members direct and command; we very seldom ask. It is bred in us from basic training and instilled in us at the academy. It is conceived right around the time the drill instructors are mentally breaking you down only to build you back up as a warfighter, a machine that does not question but only reacts to the directive given. Think about it: your computer does not question why you punch certain letters on the keyboard. It simply does what it is told. Of course, we are humans and there will always be some questioning attitude that must take place and within reason. But the idea is to ensure the service member does not hesitate when given an order; they simply carry it out.

If we are honest, service men and women tend to view some superiors that are nice and kind as soft or as someone who will not act swiftly to correct or discipline when an order is not followed. It goes unsaid, but make no mistake about it, it is understood by all that they cannot oversee their division, platoon, battalion, or department without being run over from the least to the greatest. Giving orders reinforces the rank structure. If you have the authority to give an order, that automatically puts you in a category that demands respect of your position or your rank whether you are liked as a person or not.

> When we speak sharply to our children as if we expect them to carry out orders and execute them to perfection, it damages the way they may view us and hinders our relationship with them.

The problem is when we come home, our kids do not know or respect us as a chief, gunny, lieutenant, sergeant, or petty officer. They just want their mommy or their daddy.

The problem is when we come home, our kids do not know or respect us as a chief, gunny, lieutenant, sergeant, or petty officer. They just want their mommy or their daddy.

So, when we speak sharply to our children as if we expect them to carry out orders and execute them to perfection, it damages the way they may view us and hinders our relationship with them as time goes by. This is not the case for every service member, but we have all made this mistake at some point regardless of MOS, rating, or job title we serve in. Because of serving in some of those more dangerous and life-threating jobs, the negative impact on our children tends to be more evident. Such may be the case for those serving as law enforcement, detectives investigating crime scenes, firefighters, and paramedics. The need for protocol and following of orders is essential to the success and safety of those who perform these services. There is no doubt it has affected the way they interact with their families as well.

Nurture vs Neglect

ASIDE FROM THE WAY we talk to our children, just the fact that we are gone for lengthy periods of time on a consistent basis can plunge our children right into emotional issues that we never intended to happen. For me, it happened on the morning I was scheduled to get underway. One thing I always did before I went out to sea was wake my kids up early in the morning and tell them good-bye before I left to go to the ship. On this morning, when I woke up my son and told him daddy had to go back out to sea, he said something that blew me away. I didn't expect his little four-year-old mind to formulate the words that came from his mouth. He said, "I hate you! You leave me!" I was broken. What was I supposed to do with that? What do you say to that? I was at a loss for words. I wanted to be upset with him for the way he was being blatantly honest, but I was more upset because he was right. I did always leave him. There were many times he wanted me and needed me, but I was not there. It is at this point that damage is evidently inflicted upon both parent and child. The child feels neglected and unloved because they need their parent, and the service member feels overwhelmingly guilty for leaving behind a child they want to nurture. We repeat this cycle over and over to support and take care of the ones we love. Yet, it can appear to them as if the way we love them is by leaving them.

As parents, God charges us to nurture our children, raise them with morals, and teach them to walk in righteousness after God. According to Ephesians 6:4, we are to do this without provoking our children.

"Fathers, do not provoke your children to anger, but bring them up in the discipline and instruction of the Lord" —Ephesians 6:4, NLT

In this passage, Paul is speaking to fathers on how to operate in their role within the family, but this idea of not provoking the child holds true for both men and women. To *provoke* means to aggravate, irritate, or enflame. Anybody who has children knows it is impossible to not aggravate or irritate a teenager and telling a two-year-old they cannot have

> The child feels neglected and unloved because they need their parent, and the service member feels overwhelmingly guilty for leaving behind a child they want to nurture.

candy will sometimes enflame them to throw a tantrum. So, how are we to find a way to raise children without doing those things? Let me explain. In raising children, of course, there will be times they are aggravated with us because we may not approve of some of their friends. Yes, they will be irritated because we will not let them eat ice cream for dinner, and yes, they will be infuriated or enflamed when we take away their phones after rummaging through their social media accounts. That is not what I am talking about because it is done within a reasonable means to protect and ensure their health and safety.

What I am talking about is what the Bible instructs us to—raise and discipline our children without creating an environment where they are constantly upset. Aggravation and irritation, in this instance, comes when a child is constantly being berated and belittled because they are made to feel as if they can do nothing right. The atmosphere we raise our children in is enormously important to their development. Because of the environment in which we work and the cold, sarcastic way in which we communicate, service members tend to provoke our children without even knowing it. We can provoke our children when we talk *at* them instead of talking *to* them. This is what my wife was talking about when she would say, "The children are not your sailors!" I did not realize how I was neglecting my kids.

You may be thinking, *I do not neglect my children; I give them everything they need.* I am not questioning your ability to provide food, clothing, and shelter. I am not talking about that kind of neglect; I am talking about neglecting your child's feelings and emotional needs as well. I did this too often as I would not consider how my son or my daughters were feeling. All I could see were their mistakes, like writing down a list of "hits" or discrepancies found during a uniform inspection. You are never told how good the rest of your uniform looks, only that one ribbon is out of place, that your cover is crooked, or you have a stray thread the Navy calls an

"Irish Pennant" hanging off your uniform. All I could see was homework that wasn't finished and a room that wasn't clean. I was trained to identify mistakes and to find deficiencies in the most minute details. I did not except excuses for why homework was not completed, I did not want to hear why the room was not clean; the order was given and the expectation was that it would be carried out swiftly to completion and without error.

So, instead of nurturing my children, I neglected the ten things they did right to point out one small mistake I found. I neglected my children's emotions because it got in the way of them doing what I told them to do. I thought they would use their whining as a way of getting out of doing their chores to the perfect standard I had set for them. But they were showing out because they wanted more time with me.

When I came back from a three-week underway at sea and pulled up to the house, my fifteen-year-old son came trotting down the driveway with a big smile on his face

> Military parents love their children, but sadly, we tend to give them orders to carry out instead of encouragement to carry on.

to meet me. However, to his dismay, I vehemently voiced my displeasure at the way he cut the grass. I looked right passed the glistening smile of my child to see the sun scorched grass on my front lawn. Instead of praising a fifteen-year-old for cutting the grass on his own, I yelled at him for cutting the grass too low. I didn't realize until after I said it, but instantly the look on his face clearly indicated he was deflated, heartbroken and devastated. Not only did I provoke him, but I also neglected him of time he was looking forward to in reconnecting with his dad. Now instead of spending time with the one he has been waiting to see, he was secretly wishing I would just get back in the car and leave again. Military parents love their children, but sadly, we tend to give them orders to carry out instead of encouragement to carry on.

Blended Families

MANY MILITARY FAMILIES ARE blended families consisting of children from previous marriages or relationships blended into the current marriage. Stepchildren can have a tough time transitioning. They must try to adapt to a new parent figure, an unfamiliar environment, and if they are not acquainted with the military, they have to learn to adapt to that as well. It is imperative to remain calm and allow the child to express themselves in a respectful manner. Any discipline actions initially should continue to be by the biological parents if they are active in their lives. Remember, children who express, "You are not my dad!" or "You can't tell me what to do because you're not my mom!" are really expressing their sadness because their biological parent is not there. This transitional period can be tremendously difficult for children and must be managed with extreme care and compassion.

Instead of getting angry with the child, try to sympathize with them. It is hard for a child to understand this process and will often blame themselves for the separation of their parents. In addition, they may also view you as an obstacle preventing their biological parent from being together. At any rate, even if you are a stepparent, you are still a parent and are obligated to nurture, encourage, develop, and raise them to become happy, successful, and confident people. You must make it your mission to refrain from neglecting, ignoring, or disregarding their needs, whether physical, spiritual, or emotional. This takes a little skill and a lot of effort because time is limited if you are in a training phase that puts you in the field or at sea for weeks at a time. It also requires you to deconflict with any children you may be bringing into the marriage, so they do not feel neglected. It is hard enough to share a parent with the military; sharing a parent with a new family can be hard as well.

Let Children be Children

Too often we try to give children adult roles and positions to fulfill to replace your responsibilities while you are gone. Don't tell your little boy it is his responsibility to take care of his mother and the house while you are gone. It is common amongst military and civilian families alike that when a father is gone, or is about to leave, he or the mother instantly and without any thought as to what they are saying, anoint the oldest son as head of the family by telling him he's the man of the house. This means he is now responsible for taking care of the family in the absence of his father. This makes logical sense that the only male in the house must now assume the role of the other male who is now gone. Yet, no one considers that this boy is not a man, so there's no way he can uphold the role. Listen, we have all done this. I have and so did my father, but do not give your boy a man's responsibility to uphold when he should be focused on other things. You have just put a child in an adult's role and added unnecessary stress they should are not yet equipped to endure.

The same can be true when military moms tell their daughters to "Take care of daddy and watch over the house." This seems unsettling to some, but I have heard military moms say they have told their daughters this very thing. The idea is that in giving the child your responsibility to uphold,

"It must be noted children are not psychologically developed or mature enough to give advice, to soothe a grieving parent, or add emotional support."

it somehow keeps them connected to you by doing what you do. But a daughter is not a woman and cannot tend to the needs of the family the way a woman can. In both these instances, it must be noted children are not psychologically developed or mature enough to give advice, to soothe a grieving parent, or add emotional support. Children have their own struggles in dealing with not having their parent around that often leads to a form of depression or anxiety.

I remember when my dad's job moved him to Memphis, he told me to be the head of the household while he was gone. I was going into my freshman year of high school, and I remember on one hand feeling like I was free to do whatever I wanted to because he would not be here to stop me, but on the other hand, I also remember feeling alone. I took on a man's responsibility and thought if something happened to my mother or my sister while he was gone it would be my fault. I was not equipped to raise a young girl; I was still a kid myself. I had no idea how to comfort my mother who was struggling with my dad being gone—I was inadequate in that department. The stress of that time in my life manifested itself in some outlandish teenage behavior.

Military children tend to develop a sense of independence early on, and as they grow and their parents deploy more often, they will struggle to switch between child and parent roles, until eventually, the child and the returning parent bump heads. Part of this independence is developed out of necessity because their parent may be on deployment or in the field and they learn to do more things by themselves. However, some of it is a result of the roles and responsibilities we place on the child for which they are not ready. The best thing to do when the active-duty parent is gone, is to simply encourage them to be good, have fun, and when asked, help around the house. This allows them the opportunity to aid the parent that is at home without the burden and responsibility for any negative outcomes.

> The best thing to do when the active-duty parent is gone, is to simply encourage them to be good, have fun, and when asked, help around the house.

Application:
Children have voices that are oftentimes silenced by their parents or guardian. Communicating with our children aids them in learning to communicate as adults.

- Sit with your child, take them for ice cream, or to their favorite restaurant and ask them their true feelings about your career. It is important to hear them and to let them feel heard. Help your children identify their emotions if they get upset. Talk them through the reason behind their feelings.
- If they get angry, scream, slam doors or throw objects, you must help them identify why they are behaving negatively and give them something positive to replace those behaviors.

Chapter 7
Is the Ship Your
Side Piece?

"Being gone and being out to sea so much left me feeling uncomfortable when I was home, like I didn't fit in with my own family. So, I started a new family. The Navy became my "mistress" and the sailors under me became my "kids.""

THIS NEXT TOPIC IS a touchy one. It deals with one of the most important aspects of any relationship—time. Time is valuable; it's something we can never get back. Being in the military we spend a great deal of time away from our loved ones, family, and friends. I feel this book will not only bless those who have been married for years but will benefit any military person thinking about getting married. It is with these young couples in mind that I came up with the title for this chapter. For those of you not familiar with the vernacular of modern slang, let me elaborate. I am not talking about a pistol that is strapped to your thigh in the event you must put down your rifle on the battlefield and go to a backup weapon. But when a person who is in a monogamous relationship becomes romantically involved with someone else, but does not leave their current love, the new person is called the *side piece*. They are not the main attraction or the major object of affection; they are the one on the side. They are the backup just in case things start to go badly and you run out of ammunition. While this may be a rough analogy of romantic relationships, it is imperative to understand there should never be a side piece when speaking in terms of marriage.

We are all familiar with people who have not been faithful in their relationship with their spouses or significant others. When this occurs, it causes the spouse—the main attraction—the one who is supposed to be the object of your affection to feel as if he or she is not enough. Feelings of inadequacy begin to rise and emotions run amuck as they feel there is something wrong with them and you no longer hold them in your heart the way you once did. If you were happy with them, you would not be spending time with someone else. In many cases, the military branch we belong to inadvertently becomes the infamous side piece. Always there, waiting in the background, knowing your heart is with your family but steadily demanding more of your time.

The side piece is the side piece because there is a contribution being made to you by giving you something you desire. In respect to the military, it is in the form of money, power, and other benefits you feel you need. Therefore, just like the relationship side piece, you now become obligated to keep them around because they are providing a need and contributing to a certain lifestyle. A need

> In many cases, the military branch we belong to inadvertently becomes the infamous side piece. Always there, waiting in the background, knowing your heart is with your family but steadily demanding more of your time.

you have become so accustomed to getting, you refuse to accept the possibility to live without it. To be successful in the military, you must develop a love for it, and you find yourself caring more than you should about the job or the mission you are tasked with completing. You invest your time: ten, fifteen, and sometimes twenty years or more. If not careful, the love, time, and devotion you show to the branch of service you are in, will make your spouse feel as if they cannot compete with your job.

I remember my wife told me I had been having an affair for fifteen years. I looked at her like she had two heads, but she was right. I was married to her but spent most of my time with the Navy. While my family was at home eating Christmas dinner, I stayed the night with my side piece. You see, many times I had to stand duty, and in the Navy that requires you stay onboard the ship a full twenty-four hours. Depending on what ship platform you are on, that duty day could roll around every eight days, every six days, or if you are on a small ship, every three or four days. God forbid you are a part of a major evolution and find yourself in "Port and Starboard" duty. Then you are onboard for twenty-four hours every other day, and this is when you are not even out to sea. I spent several birthdays with the Navy. I took trips around the world, built memories, and experienced new things with the Navy I never had the opportunity to experience with my own wife. The Navy became my side piece.

If it is one thing I have learned about women, it is when they are with you, they do not like to share their time with anyone else. They may accept it to some degree, if they do not feel it is threatening to the relationship. They will even make excuses for why it's this way. In the military, the reason things are the way they are, go beyond defending our country or fighting for freedom. One of the reasons why military spouses accept less time is because you cannot simply quit and work somewhere else like a civilian can. A military career affords benefits and retirement opportunities most companies cannot compete with. Additionally, a military career compared to most conventional careers is extremely short lived. At twenty years in the military, you can retire from service and immediately draw a pension. Therefore, military spouses may make the excuse that although it's bad now, in a few years it will be much better because the service member retires, and we can spend as much time together as we want. Unfortunately, for some, that time does not come because the side piece has become so demanding of your time, talent, and attention that it become number one and takes the place of your spouse. Hence, the moment a wife feels your career is affecting their relationship with you, they will come to resent it.

This dynamic is not military specific. I believe this applies to anybody in a relationship, whether you are a doctor, lawyer, chef, or grocery store clerk. If your organization increases its demands of your time, you can bet there will be tension at home if your personal time begins to diminish. Many pastors and clergy members unintentionally fall victim to this situation as well. They are passionate about their spiritual calling, they are driven by their commission, and as a result, they put all their time and effort into trying to do what the Lord has told them. Just like any other profession, there must be precautions taken to prevent the erosion of the marriage. The reason there is such fallout and backlash over a pastor that has fallen into an unduly relationship with another person outside of their marriage is because they are held to a higher standard. More times than not, it was the ministry that became the sidepiece long before the mistress did. The relationship grows stale and unfulfilling because they are engaging more with the church, the elders, and the saints than they are with their own spouse. As the old saying goes, "The road to hell is paved with good intentions."

On the other hand, some men like to feel needed. Personally, it gives me a sense of purpose, it makes me feel resourceful, and depended upon. It caters to the inner part of a man's makeup to be protector and provider. When a woman tells her man thank you for doing something or for helping her with a task she physically cannot do, it wakes something up on the inside for the man to do the best job he can to make her happy and continue to depend on him for protection and provision. Some men do their best work when they are needed. I have heard it in the huddle during football games when the coach says, "We need you on this one" or during a timeout in a basketball game, "We need you to hit this shot" or at work in the corporate office, "Close the deal and get this job done; we need this." There is something that motivates a man when he is needed. It validates our purpose, it grows our confidence, and if our lady expresses she needs us, it makes us feel important in the relationship.

As service men and women, we must constantly be on guard to not allow the battalion to steal all our time. We cannot let the ship preoccupy our thoughts even when we are not there. Also, we must not allow our jobs or branches of service to make us feel as if we are needed there more than in our families. Oh, sure, the demand will always be there, the responsibilities will always be there; however, we must learn to take the uniform off when we come home. We must learn to separate as best we can. Yes, service members are technically on call 24/7, and we may have to leave in a moment's notice, but it is how you manage the moments before the notice that will dictate the success in your relationships.

"

There is something that motivates a man when he is needed. It validates our purpose, it grows our confidence, and if our lady expresses she needs us, it makes us feel important in the relationship.

"

Reassurance

I DO NOT BELIEVE anybody seeks to be unsuccessful or to purposely fail at the things they do. Nobody wakes up and thinks, *Today I am really going to blow it at work*, or as soon as they get married, *I can't wait to drive this marriage right off the cliff*. Unfortunately, without proper communication and growing together as a couple, that is exactly what happens to men and women across our nation, and those statistics are even higher for those serving our country in the military. As I was working to climb the naval ladder of success and advance within the ranks of my community, I was neglecting my family in the process. I kept telling myself when I get orders to a different location, it will be better; when I reach the next pay grade, it will get better; and when I assume a top position, it will get better. However, I am here to tell you that when you focus most of your efforts on your work and expect the success of your hands to make your family happy, "better" never comes. You may make more money, but money does not equate to happiness. What good is going on a trip or taking a cruise if the people you are going with can't stand to be around you. We are so trained to reach lofty goals, that we

> We are so trained to reach lofty goals, that we fail to realize the little achievements are what develop strong relationships.

fail to realize the little achievements are what develop strong relationships. Going to a child's softball game or a dance recital no matter how tired you may be develops strong relationships, because it shows them you are available. You invested time into them instead of just throwing money at them. Your kids don't need another pair of Jordan's for their feet. They don't need another phone or video game. What they need and crave is your attention—time that comes from an extremely busy parent willing to give that extra time to the development of their children. This is the best way to reassure your children they are more important than your job.

Your spouse must be reassured as well; your husband doesn't need a motorcycle, he needs you. Now before I get ahead of myself, I think I speak for any rider when I say a motorcycle is a great gift, but it means nothing if his wife does not express her thankfulness and encourages him to continue on with her support. Your wife doesn't need another purse or a pair of red bottom shoes. While she

> What they need and crave is your attention—time that comes from an extremely busy parent willing to give that extra time to the development of their children.

may be grateful for the gift, it will never take the place of her husband reassuring how much he loves and wants her in his life by emphasizing her importance over his occupation. I must also add that when you are seeking to reassure your partner it is imperative that you are showing love in a way they enjoy being loved. If you try to give a gift to your wife but what truly makes her happy is quality time, you are still missing the mark. Do not reassure your loved one the way you would like to be reassured, but rather, do it in a manner they would like. Reassuring one another in a relationship is key to guaranteeing the health of that relationship. This can only be done if you know what acts of love your better half enjoys. There is a terrific book called *The 5 Love Languages* by Gary Chapman that speaks to this topic. Countless authors who have referenced it in their own work is a testament of its value to a successful relationship. Chapman has also authored *The 5 Love Languages Military Edition* that service members can better relate to.

What Does it Look Like?

REASSURING YOUR FAMILY OF their importance and the significance you place on their value is imperative, but exactly how do we do that? One way is by leaving work at work. This is not just taking off the uniform, this is making a concerted effort to refrain from discharging all your problems from work on your family. My wife can tell you the meaning of naval acronyms, operating procedures, and military regulations, but she has never served. She got all that info from me. It is not bad to discuss your job, in fact, it is good. But leaving work at work is also for the health of the service member. We tend to talk about things we have coming up and the things we must do for the next certification or the next exercise; we go on and on about the training we must attend and the meetings we have next week. Your home is supposed to be your haven—your sanctuary. When you do not actively disengage, you never detach from the job, and you keep the stress in your body.

I would suggest that you also make your family a part of your career decisions. I have made it my business in the later part of my career to include my wife in every decision regarding my orders. I remember in the beginning, I did not; I was extremely selfish, and I thought, *If I am the one going out to sea or serving at any shore installation, then I am making the decision. I am the one that's bleeding, sweating, and putting their life on the line.* This is not, I repeat, this is not a good idea. I uprooted my family from San Diego, California, and moved them all the way to the other side of the country to Virginia Beach, Virginia, without asking how they felt about it. I just told my wife I had orders and when we were leaving.

They were so upset; they loved living in California, but I was more concerned about what I wanted. All these things must be taken into account when making military decisions. They never let me off the hook, and often reminded me of it even up to ten years after the move. It would have been better to inform them of the possible orders, discuss the pros and cons, and ask their opinion. Even if I took the same orders, I may have left California with their support instead of their resentment.

Finally, keep the family well informed with troop movement. As we all know, these schedules change at a moment's notice, but the more you communicate the schedule, the better you and the family can prepare for separation as well as plan much needed family events like leave and vacations.

Application:
- Include your family in the decision-making process for orders you will be taking. This shows you value their input and incorporates them in your career.
- Ask your family to attend any command events such as cookouts, award ceremonies, parties, or balls.
- Bring them to your job and show them where you work. Introduce them to your peers; this shows a spouse you are not ashamed of them or "hiding" them from others.

Chapter 8
General Quarters,
General Quarters

"I have never failed at anything I have ever done, but now my marriage was failing, and I didn't know how to deal with that."

From the time I was in boot camp, through the now twenty-four years of active-duty service, there is one phrase that when heard across the 1MC (the ship's intercom system), automatically shifts my mind to prepare for war: "General Quarters, General Quarters, All Hands Man Your Battle Stations!" It is a call that signals the entire crew to stop whatever they are doing, whether eating, sleeping, running on the treadmill, or participating in a religious worship service, and report directly and immediately to their battle stations. A battle station is a place you are required to go based on your job or your qualifications that require you to support the ship in combat while at sea. Many times, you may be called to general quarters even though the ship has not been fired upon mainly if the potential for a combative exchange is present.

There is another aspect of General Quarters and that is the ability of the crew to perform damage control. Damage control is managing damages or casualties inflicted due to mechanical failure, accident, or acts of war, and it could require shipboard personnel to fight fires and stop flooding.

Fires are extremely dangerous to a ship, and if not dealt with quickly, they can spread awfully fast severely crippling the rest of the ship. Fires burn living spaces, ignite fuel or weapons, and incinerate hazardous materials that can lead to the death of many sailors and even prevent a ship from being able to defend itself in battle. Flooding is another major casualty that may lend itself to the demise of sailor and ship. Damage control men and women must act quickly to plug or shore up holes in the ship's hull from explosions to prevent the ship from sinking.

Many are familiar with the terrorist attack on the USS Cole while in port at Yemen. A small boat with explosives was driven directly into the side of the ship and exploded, leading to the death of seventeen US Navy sailors. However, it could have resulted in the complete loss of the ship had the crew not responded to extinguish the fire and secure the spaces affected by flooding from a twenty-foot hole in the side of the ship. Similarly, there are many times your marriage will come under attack; you must sound the alarm and respond to General Quarters when you notice a threat.

Battle Stations

ONE OF THE MOST taxing aspects of being on a ship in hostile waters is constantly being at battle stations for extended periods of time. General Quarters is not just for combat. You also go to general quarters when there is the possibility or threat that could potentially lead to battle. As military spouses, we are placed in situations on a consistent basis that are potential threats to our marriage. We work in such close proximity with one another and for such prolonged periods of time that the potential for an affair is ever present. Not all military spouses are cheating or doing something wrong; quite the contrary. But because so much time is spent together, working relationships must be formed to complete the mission. Those working relationships turn friendly as we strive to create an environment conducive to productivity. Where we can get into trouble is when we fail to acknowledge or disregard conversations that may arise which we know our significant other would not appreciate us having. Let me just say it this way: if you find yourself in a conversation with someone, and you know you would not have that same conversation if your spouse or significant other was present, you need to immediately go to General Quarters. Most conversations are harmless, and that is how they always start, but by being "on guard," you can identify situations before they arise and set yourself to battle stations.

You may be thinking this sounds a little extreme, but it is much easier to be prepared and divert an attack than trying to do damage control after a torpedo from an adulterous relationship has been fired at your marriage. These attacks are easy to divert when your marriage is healthy, but if you have grown apart and have stopped having conversations with depth and substance, stopped going on dates together, and frankly no longer "feel" in love, your marriage has become what we call a "soft target."

I must also say the service member is not the only one susceptible to this. The spouse left at home for lengthy periods of time must endure the same temptations and be ever vigilant. We have all heard the horror stories of that one person who comes home after a deployment to find the house empty and the bank account depleted. Or the one who receives a letter or email while on deployment that reads "I

"If you find yourself in a conversation with someone, and you know you would not have that same conversation if your spouse or significant other was present, you need to immediately go to General Quarters."

want a divorce." I have even heard of women who came home and found out their husbands were living with someone else and raising another women's child while sending the service member's child(ren) to stay with grandparents. These are horror stories of extreme cases, but if there has been a breach of the marital contract, damage control must be performed.

Damage Control

DAMAGE CONTROL IS A term used to aggressively take measures to combat the casualty usually because of fire or flooding that could lead to the loss of the ship. The initial response is to put out the fire, stop the flooding, or stop whatever is causing the immediate damage to the ship and maintain as many of the systems as possible to allow the ship to function until it can return to port for repairs. In most cases, especially with couples who have been together for some time, they want the relationship to work, but they don't know how to fix it. Please understand that because of my faith, I will never promote divorce, but I would never tell a person to stay in an abusive relationship either. In the event of marital problems most couples encounter, the marriage can in most cases be saved, but here is the key. Both parties must want it to work.

This means both people in the relationship must enter the damage control stage with the mindset of "saving the ship." In respect to my brothers and sisters in the other branches, I used the analogy of the ship as the marriage, because like the marriage, a ship is only as good as the people in it. We live in the ship, and we transit from one place to another in the ship. We work together to make the ship what it is, or we are equally responsible for what it is not. This plays out in the marriage as we use the marriage to transit through life; it is only as good and only as rewarding as we make it. If we do not do constant maintenance on the ship, it will begin to corrode and rust; if we are not proficient in training when trouble arises,

> In the event of marital problems most couples encounter, the marriage can in most cases be saved, but here is the key. Both parties must want it to work.

we will not be able to save the ship. These things are true in our martial unions as well. So, if we do suffer a casualty to the marriage, we must set our mind to do everything to save it. Once that decision has been made there, are several steps that must take place.

1. **Put Out the Fire or Stop the Flooding:** Immediately stop the action that is causing the casualty. If it is another man or another woman, sever the relationship without delay, and never call or contact them again, even if you want to. If there is verbal, emotional, or physical abuse, acknowledge your deficiency and stop the hurtful actions. Maybe the one thing you are doing is neglecting your spouse. This could be from emotionally removing yourself and not inserting yourself back into the relationship. Whatever is causing the most pain, stop that action first and refrain from doing it again.

2. **Assess the Damage:** This one is hard because it requires you to not see things how you want to, but to see them as they are. The damage assessment of a ship is critical to determining the ship's health and how capable it is to defend itself or carry out the mission. A dishonest assessment may lead to believing the ship is in better shape than it is, putting it and the crew at a disadvantage.

Is it possible for the ship to be fixed, because if it is beyond repair, the ship may be abandoned by the crew, scuttled, or taken out of service. These same type of questions must be asked when assessing the marriage. How bad did you hurt your spouse? What is your role in the casualty and how can it be fixed? We always like to blame the other person, but you must be honest with yourself and take responsibility for your own shortcomings.

3. **Devise a Plan for Repair and See it Through to Completion.** The ship will undergo a maintenance phase to address major concerns, make repairs, and install upgraded equipment. Anybody that has experienced a maintenance availability onboard a ship will tell you the work to be done is meticulously thought out, planned, and tracked to ensure everything is completed. A maintenance availability is some of the most strenuous environments that require dedication, determination, and major work. This same dedication, determination, and major work may be required for a marriage in need of major repairs. The plan may require counseling. Too many people, especially men, do not want to go to counseling. It is because we do not like opening up and feeling vulnerable to strangers. We think we can fix it, but trust me, not talking about it and acting like everything is okay will not fix it. Your plan may include a deeper devotion to your faith, being intentional in gift giving, conversations, and outings. Whatever your plan is, it must be agreed upon by both people to work at it together and maintain honesty at all times. Be patient and understand that your marriage is worth saving regardless of how you may feel now. The person you fell in love with is still there; you must find them, connect with them again, and work to fix the relationship.

"

Be patient and understand that your marriage is worth saving regardless of how you may feel now. The person you fell in love with is still there; you must find them, connect with them again, and work to fix the relationship.

"

4. **Get Back in the Fight.** There is a little jingle that goes, *"Sailors belong on ships, ships belong at sea. Haze grey underway, that's the life for me."* There is truth to this: ships belong at sea not tied to the pier under constant repairs. A ship that cannot get underway is of no use to the Navy. You need to understand your marriage needs to be in proper operation; it needs to be functioning the way God intended to get the most out of the relationship. Your marriage is valuable; it is designed to bring pleasure, happiness, and security to your life. Do the work to get your marriage ship shape and get back in the fight!

Application:

- What is causing the most pain in your marriage and how can you put a stop to it?

- Do a self-assessment of your marriage and your role in it.

  o Determine what must happen to repair what is broken.

  o List what you can honestly work on yourself.

  o List what items, if any, you need to seek coaching or professional help for.

- Getting back in the fight is critical. Make a vow with your spouse that you are willing to fight for your marriage!

Chapter 9
Darken Ship

"And God called the light Day, and the darkness he called Night. And the evening and the morning were the first day" —Genesis 1:3

THERE IS NOTHING MORE distinct than the separation between light and darkness. Aside from the metaphors of good and evil, light and darkness are periods that are necessary yet serve different purposes. During the day, we go through the hustle and bustle of our lives to get things accomplished. Children get on loud school buses filled with chatter and are whisked off to school during the day. We go to grocery stores to get food for Sunday dinner and visit nail salons and barber shops during the day. It is during the daylight hours most of the world's business takes place, because naturally, nighttime is when most of humanity is tucked in their beds fast asleep. It is for this reason, that things that are meant to be kept secret seem to happen at night. It is under the cloak of darkness many military missions have taken place making the element of surprise much more effective. It is harder to see things at night, details are obscured, and identifying markers are concealed. Everyday ships throughout the fleet mark the transition from day to night by the command "darken ship" given over the 1MC.

Darken ship is a condition where all the ship's external lights are turned off and all outside hatches are secured in a manner to prevent any internal light from being seen from outside. Inside the ship, fluorescent lighting is extinguished, and red lights are turned on in their place. These precautions allow for the ship to remain hidden or concealed from others as it transits through the sea. I can't help but draw the analogy of a ship in concealment to the possible mental state of a sailor onboard. Just like darken ship is an intentional posture that must be entered, many sailors intentionally enter a darkened ship state of mind. They hide and conceal their real feelings— feelings of loneliness, depression, or sadness, and quietly transit the lonely ocean of their lives as if they are the only ship in the world that is at sea. This is the sailor's attempt to deal with their problems alone without being detected by their chain of command, family, or friends.

The United States military has tried to combat the stigma of a person struggling with mental health having it held against them during evaluation time. I have witnessed the winds of change rapidly blowing during my years at sea, and I have taken notice of the difference in generations. While so many younger sailors coming into the Navy are more open to expressing their need for mental health, older service members are apprehensive about mentioning their struggles. I believe mental health issues have always been there; they have just been handled differently throughout the generations. However, regardless of the time served, many troops still believe getting help for any mental illness is a scarlet letter that will negatively impact their career. Therefore, sailors as well as marines, soldiers, and airmen turn off the lights to their visible problems and mentally go to "darken ship" to evade detection.

Darkness Falls

I AM GUILTY OF attempting to conceal my problems and the effects it was having on me mentally. I have been in some dark places during my career—places that left me feeling alone, broken, and ashamed because of the decisions I made. I have lost close friends that I served with that later would prevent me from forming close relationships with other people. I have felt estranged from my parents and siblings because I rarely went home. I single handedly dismantled my wife's feelings and tormented her spirit with my actions and my words. I watched my son become a stranger and both my daughters withdraw from my embrace with little emotional expression. I have been in the dark waters of depression; I have been engulfed with anxiety without understanding what was happening to me. Yet, I chose to press forward. I believed that by attempting to stay the course I could prevent people from discovering how dark it had gotten in my life. My over achievement during this time would not give any inkling that I was struggling, that my marriage was full of holes taking on water, and that my purpose for life appeared grim at best.

I will never forget the day I was pinned as a Chief Petty Officer. I was selected, initiated, and accepted into the hallowed halls of the Chief's Mess. I endured many years of demanding work, separation, and sea time, and I had finally reached the pinnacle of my career. It quickly dawned on me as those prestigious collar devices of the gold-fouled anchor, with the silver, superimposed USN were being pinned on my uniform, that this did not fulfill me the way I expected. I had reached the upper echelon of the enlisted ranks, and I was miserable.

The joy of being mentioned amongst this august body of people was soured by the tornado forming in my personal life. I quickly discovered my new position, the extra money, and the added respect did not make my life better. So, I did what I always did—I put my head down and started grinding. I made it my mission to outwork everyone around me by obtaining qualifications that set me apart from my peers. It was, however, a failed attempt to feel better, to find some worth within myself that I could no longer see. It felt like my life was crumbling all around. I was watching everything I had worked to build falling on top of me in slow motion, and I had no way of escaping. So, I chose to turn the lights out.

Taps, Taps

TAPS IS A TIME-HONORED tradition that had its beginnings on the battlefield of the Civil War. The twenty-four note tune we refer to as "Taps" was created in 1862 after U.S. General Daniel Butterfield was dissatisfied with the current bugle call to signal to troops it was time to go to sleep. Not long after the creation of this new bugle call, it was utilized at a funeral for a Union cannoneer killed in action. This was thought to be safer than the original firing of three rifle shots so the nearby enemy would not confuse it with an attack. Today these same twenty-four notes are played to commemorate the memory of members of all branches of the armed forces.[2] Taps is also played on military bases to signify the start of quiet hours. For ships at sea, a simple command: "Taps, Taps. Lights out. All hands turn into your racks. Maintain silence about the decks."

The word "sleep" is often used as a metaphor for death as the person may appear to be in a state of deep sleep.

2 https://www.history.com/news/how-did-taps-originate

It is one reason the bugle tune signaling the troops to turn out the lights and go to sleep was so widely adapted for military funeral service. Sleep denotes death. Onboard ship, an enlisted sailor's living and sleeping quarters are referred to as a berthing. Before I was commissioned as a naval officer basking in the comforts of a two-man stateroom, with comfortable beds and plenty of space to relax, I was a sailor who humbly lived in the cramped, musty smelling, overcrowded enlisted berthing of an aircraft carrier. The interesting thing about the ship's berthings is the morbid name given to the beds sailors sleep on: "coffin racks." This name is given because the top part of the rack swings up like a door, and most of your belongings are tightly packed inside its aluminum frame. The top is lowered back down and the sailor sleeps on top of their belongings. The call for 'darken ship" only turns out the lights in the main passageways, but the order of "taps" requires all lights in the berthings to be turned off. I can't help but draw the connection to the fact that when taps is ordered and the lights are turned out, the sailors sleep in coffins.

There is a starting point to the darkness that consumes us. We may not initially know when that is but once darkness descends in the mind, it is only a matter of time before taps. I was in a place many service members find themselves in—a place in my life so dark that I was ready to end it all. One evening, I got in my truck and slowly drove down the shadowy open road before making a right turn into the parking lot of the church we were attending. As thoughts of despair and feelings of hopelessness consumed me, I pulled my pistol from the center console of my truck. Feeling the cold steel in my hand, I gently placed it on my lap. Letting out a huge sigh, I glanced at the clock on my dashboard. The tears in my eyes made it hard to decipher the numbers, but as my vision became more clearly, I was able to see the time. Wouldn't you know, it was almost taps. Here I am, a man of faith. I am saved, sanctified, filled with the Holy Ghost, but I felt so far gone that I slowly raised the gun placing the barrel in my mouth. I was ready to turn out the lights and go to sleep.

Then There was Light

WITH THE RISING OF the sun comes the dawn of a new day, a fresh start, and new opportunities to be had. Light is essential to life; it illuminates, reveals, gives warmth, and provides energy. Light allows us to see our way when it is dark. On that night, that is exactly what happened. As bad as things were, I could not help but think about the consequences resulting from the pull of that trigger. My wife would be destroyed; she would be without her husband, her protector, and provider. As toxic as our relationship was, it was still ours, and losing what was left would send her to a place she would never recover from. My children would struggle through life with no mentor, no father figure, and no dad to praise them for their accomplishments and forgive them when they make mistakes. My parents would have to suffer through the funeral services and try to comprehend the logic of having to bury a child. There would be no wake or viewing. This would be a closed casket, leaving my family and friends trying to hold on to their last image of me in their memories.

The powerful thing about being in a dark room, even if the door is closed, light always shines through the crack at the bottom. I was at my bottom, and just as God spoke

I was at my bottom, and just as God spoke into the darkness and created light, it seemed as if He was now speaking into the darkness of my situation.

into the darkness and created light, it seemed as if He was now speaking into the darkness of my situation. The Lord spoke to me on that night, reminding me of everybody that would be affected and of everything I still had to live for. I was called for much greater. The light of that truth began to shine bright in my soul. I removed the gun from my mouth, put it back in the center console, and drove home telling no one about the events of that night.

Don't Wait

IN THE GOSPEL OF John chapter 8, Jesus begins to teach, and it is recorded that He says,

> "I am the light of the world. ... Whoever follows me will never walk in darkness but will have the light of life." – John 8:12, NIV

Part of me says I was too scared to go through with it, and that may be true, but I also know that on that night, the Light of Life presented Himself to me in the darkness and revealed my value and my worth. My Lord and Savior Jesus Christ has made it His business to save me. He saved me from tragedy as a baby, He saved me from trouble as a teen, He saved me from sin in 2002, and now He was saving me from myself. It took divine intervention to stop me from doing the unthinkable, but it didn't have to. Many service members have been where I was, but tragically their decision was different from mine. Don't wait for the worst moment in your life to come and believe there will be a divine intervention, not when you can prevent it from happening in the first place.

The first thing that must be identified is that no matter how dire the circumstances may be or how you may feel, you are not alone. Do not isolate yourself from people that can help you. Pulling away and remaining isolated will allow those irrational thoughts to take hold when no one is present to speak reasonably about the situation and help you to think more clearly. Identifying someone you can trust such as a close friend, parent, chaplain, mentor, or counselor, is imperative to preventing feelings of overwhelming emotion.

One characteristic that makes great leaders is not knowing the answer to every problem but knowing where to go to get the answers when you do not have them. Therefore, you must know your resources. Every branch

> Pulling away and remaining isolated will allow those irrational thoughts to take hold when no one is present to speak reasonably about the situation and help you to think more clearly.

has an organization that offers support services to service members and their families. That list of those services and locations of the offices can be acquired through your chain of command. One form of counseling that has become increasingly popular because of COVID-19 is "Tele Health Counseling." Some people feel more at ease talking about their problems in the comfort and familiarity of their own home than in a strange setting that may cause them to be guarded. It is also possible to make appointments through TRICARE and Military One Source; however, first, I would suggest starting with medical at your command.

There are many obstacles the service member, first responders, pastors, bankers, schoolteachers, and surgeons must face from day to day. Learning to remove yourself from the rigors of your occupation and live life to the fullest with those you love is the most rewarding experience that is achievable by all.

Application:

- Not everyone will feel isolated due to family or marital conflict; some may experience loneliness or feelings of no value due to the current phase of life problems.
- Difficulty adjusting to civilian life or retiring can make some people feel sad, anxious, and like they have lost their identity.
- Get involved; pick up a hobby or do what makes you happy.
- It is important to address these concerns immediately, and if you have suicide ideation call the Suicide National Hotline—Dial 988 from your phone or mobile device.

Chapter 10
When the Uniform Comes Off

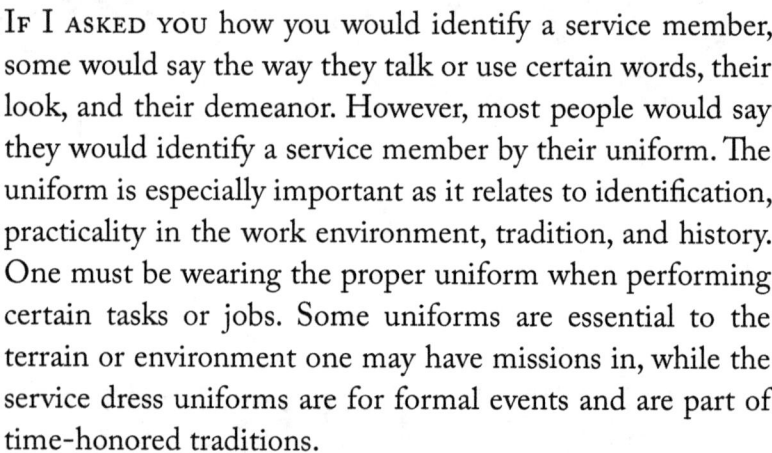

IF I ASKED YOU how you would identify a service member, some would say the way they talk or use certain words, their look, and their demeanor. However, most people would say they would identify a service member by their uniform. The uniform is especially important as it relates to identification, practicality in the work environment, tradition, and history. One must be wearing the proper uniform when performing certain tasks or jobs. Some uniforms are essential to the terrain or environment one may have missions in, while the service dress uniforms are for formal events and are part of time-honored traditions.

Yet, it is the association with the individual that wears the uniform that identifies them with something greater than themselves. Therefore, what begins to happen is the service member inadvertently attaches their identity to the uniform; this is instituted in basic training and reinforced throughout their career. First, the last name of a service member is affixed to the uniform to identify the seaman, airman, soldier, or marine. Great emphasis is placed on how good the uniform looks because it represents more than just the individual. It represents the branch of service they are affiliated with and is expected to be more than presentable—it should be impeccable.

As the service member moves through their career they will eventually transfer to another command. The first impression a supervisor has about a new military member transferring in is generated the first time they see them in their uniform. An outstanding uniform can lead to the perception that an average sailor is an excellent sailor. It can, however, also give the false perception that a particularly good soldier may be below average, simply because their uniform has not been ironed with proper creases or their shoes or boots have not been buffed to a reflective shine. To enforce the service member's close attention to detail, uniform inspections are held. If the uniform is not perfect, if ribbons and medals are not affixed with the correct measurements and placement, or if the individual is found with an unsatisfactory uniform on a consistent basis, they may be labeled a "dirt bag." A dirt bag is someone who looks unclean or unkept. It is often assumed that if you did not pay attention to detail on your uniform, chances are you will cut corners on your job and will not take the time to perform your duties properly.

We have different uniforms for different assignments, different branches, and various times of the year. On our uniforms are devices and patches that denote our unit, command affiliation, and job title. Some uniforms have stripes stitched on the sleeve above the left wrist indicating the number of years the service member has completed. All uniforms have an indication of rank which determines where you fall in the order of the chain of command. Rank determines a position and a level of respect and responsibility that is recognized by all members of the organization. We have a saying: "You may not respect the man or woman as a person, but you will respect the rank." From the moment we receive uniforms in basic training, the uniform gives us our identity amongst one another. The uniform describes and details so much about an individual that when the uniform comes off, whether that be at home after working hours, when their service obligation is complete, or after making it a career and it is time to retire, every single person struggles with their identity to some degree outside of the military. Sadly, that struggle becomes harder and harder the longer the commitment to wear the uniform continues.

Losing Yourself

HAVE YOU EVER SEEN the old sailor in the store or the old marine at the auto garage, and they have their favorite "RETIRED" hat on or some sort of service paraphernalia showing their pride and devotion to the branch they gave a substantial portion of their life to? There isn't anything wrong with that, but when they consistently make comments like "When I was a chief..." or "When I was in Korea, we...," that is a sure indication they are still stuck in the uniform they took off years ago. I am not against reminiscing; it is good to recount those stories. Some stories and memories of our comrades should never die. However, some can't seem to ever let it go because their identity is still attached to the uniform. Some spend fifteen, twenty, or even thirty years or more wearing the uniform every day. Can we really expect a two-week transitional course all service members are required to attend before discharge be enough to equip them to go back to civilian life? Is it fair to expect that it is going to teach us how to immediately take off the uniform when we leave service? Impossible! Even as I am writing this, I am reminded of how much of my identity is still connected to the uniform I am currently wearing.

For some of us, when that uniform comes off, we will lose ourselves. Therefore, veterans tend to talk about the "good old days" so much, because it allows them to still identify as a service member without physically wearing the uniform. It gives them a sense of identity because they may not feel they have one in the civilian sector. I know people that have been retired for twenty years and still refer to each other as "sarg" or "gunny." Truthfully, they lost their identity the moment they took their uniform off. There is a reason many who make the military a career possess some fear about the day they will no longer wear the uniform, even if they do not express it. That is one reason many members with extensive experience are so driven and gung-ho, because just like their name is affixed to the uniform, their identity has become affixed to the branch of service they are affiliate with. They have become the Navy, they are the Air Force, their name is synonymous with the Marines, or they have become the image of the Army's model soldier.

Losing that identity is extremely hard. It is hard to give up, but it all starts when you return home each day after work and take those boots off. It is at that point you must make a conscious decision to mentally discharge yourself from duty and stay connected to your family. You must lose yourself. You must lose the identity your branch of military service has created for you. Let me be clear, we as human beings are constantly growing, maturing, and learning. This is a result of life experiences, the relationships we develop, and the careers we engage ourselves in. There are many good things from each of these that serve purpose to enhance each person. They should be used as tools to increase our awareness of ourselves to the point we are able to become the best person we can be. We must be incredibly careful not to allow any person or anything change us to the point that we have lost who we really are and take on the identity of something else. Unfortunately, this is what happens to career service members, and it is what makes it so hard to let go of their connection to service.

> We must be incredibly careful not to allow any person or anything change us to the point that we have lost who we really are and take on the identity of something else.

Change Clothes

IT IS IMPORTANT TO remember that we wear different clothes for different occasions. For example, if you bought real estate property to flip for a profit, to save a little money during the renovation process, you may decide to do some of the work yourself. Well, you are not going to show up to the work site to hang drywall in a three-piece suit, and you also are not going to wear gym clothes to your wedding. The shorts and tank top may work well in the gym, but it is not acceptable at a wedding because it does not enhance the ceremony; it only adds as a distraction to the occasion and would only cause confusion. No matter how well your intentions may be, no matter how sincere your vows are, they will be ineffective if the groom sees you coming down the aisle in pajamas, or your bride is struggling to focus with a sweat towel draped over your shoulders and a headband staring back at her. These outfits just don't work for the occasion. So, you are wise enough to put on attire that is conducive to the event to which you are heading.

It is amazing how simple that concept is to grasp, yet we find it so hard to put into practice when taking off the military in the different areas of our lives. It works great for the mission, but it does not fit the marriage. On the other hand, it is almost unfair to expect a service member to completely remove every part of their military identity on the weekends and then put them right back on Monday morning. This is because service members do not work a typical 9-5 job; they are required to be on call twenty-four hours a day. Military personnel are not the only ones—off-duty police officers and firefighters will often report to an emergency in street clothes because they heard the call on their radios and felt obligated to report to the scene. When wearing the uniform for long periods of time, you become comfortable in them, and when you become comfortable, it is a challenge to change. It's a challenge to change because even without the uniform on you have become what the uniform represents. But understanding what items to take off and which ones to put on from your military life is detrimental to the relationships you have with the ones you love the most. Personally, I change clothes at work and no longer wear my uniform home. It is my way of transitioning my mind to better interact with my family before I get there. This helps me to mentally take off the Navy and leave it on the pier with the ship.

Take it Off

LET ME BE REAL, once you have served in the military, you can never completely take off the uniform because there are many experiences the service member encounters that are solely unique to the armed forces. They are ingrained in your soul and burned into your memory, so some things you will carry with you, however, you must know how to remove those behaviors that do not work well with your family but keep what enhances you and your relationships. It is beneficial to keep the discipline and respect for your fellow man that is

It is important to keep the perseverance, self-motivation, work ethic, and integrity that is continuously reinforced in the military member because those things are beneficial in any endeavor.

instilled in the service member. It is important to keep the perseverance, self-motivation, work ethic, and integrity that is continuously reinforced in the military member because those things are beneficial in any endeavor.

So, when I say take it off, I am saying lose that part of you that ONLY identifies with the military and nothing else. Take off the parts of your military career that do not work with the other parts of your life. The military structure of communication in the form of giving and taking orders does not work well within civilian relationships. Raising your voice and yelling at your family does not motivate them to do what you want; it only makes a bunch of noise and increases your blood pressure. Emotional detachment is needed in some degree to accomplish your military tasks, but it is not effective in marriage nor it is not compatible with tending to children. It is a cancer that will cause your family to deteriorate from the inside out before you even realize how bad it is. The inability to retain those things that enhance and remove those items that hinder, can make a spouse who has waited for you to retire after twenty to thirty years feel as if they are not enough now that your active-duty commitments are over. It can be challenging or even scary at first, but you must learn how to take off the part of you that is struggling to connect with your spouse,

> The inability to retain those things that enhance and remove those items that hinder, can make a spouse who has waited for you to retire after twenty to thirty years feel as if they are not enough now that your active-duty commitments are over.

your children, your family, and friends.

I am reminded of the story of Jesus raising Lazarus from the dead. When they arrived at the tomb where Lazarus was laid, Jesus told them that if they believed they would see the glory of God. Then John 11:40 records something interesting. It says, "Then they took away the stone from the place where the dead man was lying..." This tells me that before Jesus performed the miracle, the people had to remove the stone themselves. I am a man of faith and believe wholeheartedly in prayer, but once you have finished praying, you must do something in faith believing God will bless it. You must move the stone; you must put forth the effort in working at your marriage and then believe God to do what you cannot, and that is provide the healing. Pray and ask God to give you the strength and courage to change things you are familiar with, and to give you and your spouse the insight to identify potential problems before they arise. Finally, whether you have left service, are preparing to leave service, or are still building your career in the military, once you remove that uniform, here are a few tips to help you leave it hanging in the closet.

> Once you have finished praying, you must do something in faith believing God will bless it.

Application:
- Select several hobbies or activities you and your spouse or family like to do together.
- Set aside specific dates or times each week to "be a family" and engage in those hobbies.
- Take time to explore new interests and things that inspire you. This will allow you to learn who you are outside of your work.

ABOUT THE AUTHOR

Eric L. Allen was born and raised in Elizabethtown, KY, and was greatly influenced by both his parents, Rodney and Joyce Allen. However, it was his grandmother, Catherine Irene Decker, who as the matriarch of the family, first began to teach him about Christ.

He enlisted in the United States Navy in 1997, married his fiancé, Ellasin, and has been married to her for over twenty-four years. He is currently commissioned as a Chief Warrant Officer in the United States Navy, and throughout his career has served onboard ships on both the East and West coasts, set sail on countless underway periods, and has been forward deployed over eight times to date. Affectionately referred to as "Rev." by shipmates and service members, he has baptized many sailors while at sea, and his powerful preaching and prophetic insight have inspired many to live victoriously by applying the Word of God to their everyday lives. His ministry extends beyond the military as he currently serves as pastor of Living Water Ministries in Virginia.

He currently resides in Suffolk, VA, with his beautiful wife and ministry partner Ellasin D. Allen.

www.ingramcontent.com/pod-product-compliance
Lightning Source LLC
Chambersburg PA
CBHW060540130626
46553CB00002B/832